T0313348

Introduction to
Analog-to-Digital Converters
Principles and Circuit Implementation

RIVER PUBLISHERS SERIES IN CIRCUITS AND SYSTEMS

Series Editors:

MASSIMO ALIOTO
National University of Singapore
Singapore

KOFI MAKINWA
Delft University of Technology
The Netherlands

DENNIS SYLVESTER
University of Michigan
USA

Indexing: All books published in this series are submitted to the Web of Science Book Citation Index (BkCI), to SCOPUS, to CrossRef and to Google Scholar for evaluation and indexing.

The "River Publishers Series in Circuits & Systems" is a series of comprehensive academic and professional books which focus on theory and applications of Circuit and Systems. This includes analog and digital integrated circuits, memory technologies, system-on-chip and processor design. The series also includes books on electronic design automation and design methodology, as well as computer aided design tools.

Books published in the series include research monographs, edited volumes, handbooks and textbooks. The books provide professionals, researchers, educators, and advanced students in the field with an invaluable insight into the latest research and developments.

Topics covered in the series include, but are by no means restricted to the following:

- Analog Integrated Circuits
- Digital Integrated Circuits
- Data Converters
- Processor Architecures
- System-on-Chip
- Memory Design
- Electronic Design Automation

For a list of other books in this series, visit www.riverpublishers.com

Introduction to Analog-to-Digital Converters
Principles and Circuit Implementation

Takao Waho

Sophia University
Japan

River Publishers

Routledge
Taylor & Francis Group

LONDON AND NEW YORK

Published 2019 by River Publishers
River Publishers
Alsbjergvej 10, 9260 Gistrup, Denmark
www.riverpublishers.com

Distributed exclusively by Routledge
4 Park Square, Milton Park, Abingdon, Oxon OX14 4RN
605 Third Avenue, New York, NY 10017, USA

Introduction to Analog-to-Digital Converters Principles and Circuit Implementation / by Takao Waho.

Routledge is an imprint of the Taylor & Francis Group, an informa business

ISBN 978-87-7022-102-3 (print)

While every effort is made to provide dependable information, the publisher, authors, and editors cannot be held responsible for any errors or omissions.

To Keiko

Contents

Preface

Today, information processing is mainly carried out in the digital domain. On the other hand, physical quantities in the real world are analog (on a macroscopic scale), and a vast amount of the data transmitted between digital systems is represented as analog quantities such as voltages, currents, or electromagnetic fields. Therefore, analog-to-digital (A/D) and digital-to-analog (D/A) converters, both of which are referred to as data converters, play a critical role as interfaces between digital systems and the analog world. For example, an A/D converter translates an analog signal detected by a sensor into digital codes so that a digital processor can recognize the meaning of the sensor output. Following the processor, a D/A converter changes the digital codes into another analog signal to drive a motor, an actuator, or various physical devices. This book explains the fundamentals of data converters for students and engineers who start studying data converters.

As the world increasingly relies on digital information processing, the importance of data converters continues to increase. Data converters are now indispensable in the field of sensor networks, internet of things (IoT), robots, and automatic driving vehicles, in which various kinds of interfaces between analog and digital exist. Furthermore, artificial intelligence (AI) is built on the premise of using a large amount of information on the internet, and the broadband communication network that supports it incorporates many high-performance data converters. It is no exaggeration to say that myriads of data converters are scattered in the modern information and communication society. The more such applications become advanced, the more demanding are high-performance data converters. Their availability then paves a way to novel applications, which in turn require data converters with still higher performance. Such a dynamic positive-feedback process is expected to continue in the near future.

Research and development of data converters have progressed rapidly over the last two decades. State-of-the-art CMOS technology has a significant influence on the design and architecture of data converters. Attractive architectures have been proposed one after another, and circuit implementations have demonstrated excellent performance. Some circuits, which were once considered difficult to implement, have been reconsidered along with the advanced LSI technology. Furthermore, digitally-assisted analog circuits are introduced to data converters as a unique method that can efficiently reduce adverse effects due to imperfections of simple analog circuits. Now, data converters should be regarded as a kind of complicated system LSIs beyond the scope of conventional analog circuits.

Excellent books on data converters have been published recently. In international conferences and journals, various topics on data converters are frequently discussed. Information retrieval on the internet provides a wealth of knowledge, though it is somewhat fragmented. Nevertheless, it might not be so easy for engineers to understand the basics of data converters, because they are built on a broad technical basis ranging from transistor characteristics to circuit design and signal processing. In this book, paying attention to cutting-edge trends, I have selected and arranged relevant topics so that readers can understand the overall picture of this fascinating field. The readers assumed in this book are not only engineers who become involved in this field as newcomers but also engineers who are interested in using data converters. For those who develop their target applications by fully exploiting data converters, it is sometimes essential to understand the meaning hidden between the lines in data sheets. Of course, I hope that this will be a helpful guidebook for students by filling the gap between basic analog CMOS circuits and state-of-the-art data converters.

In writing this book, I occasionally referred to the book by Carusone et al.[1], the book by Pavan et al.[2], and the articles in the special issue on ADCs published by the IEEE[3], which included various ideas to explain essential concepts in easy-to-understand manners. I am very grateful to the authors. Without constructive discussion with many colleagues, I could not publish this book. Also, graduates from my laboratory gave me useful comments during the preparation of the manuscript. River Publishers provided me with a valuable opportunity to publish this book, and they were also involved in the editing process. I would like to express my deepest appreciation to all the people mentioned above.

Takao Waho
Tokyo
April, 2019

---

[1] Analog Integrated Circuit Design, Carusone, Johns, and K. Martin, Wiley, 2012.

[2] Understanding Delta-Sigma Data Converters, Second Ed., Pavan, Schreier, and Temes, Wiley, 2017.

[3] IEEE Solid-State Circuits Magazine, vol. 7, no. 3, Summer 2015.

List of Figures

List of Tables

List of Abbreviations

A/D	analog-to-digital
ADC	A/D converter
BG	background
BJT	bipolar junction transistor
BW	signal bandwidth
CDAC	D/A converter using capacitors
CIC	cascaded integrator-comb
CMD	common-mode voltage detector
CT	continuous-time
D/A	digital-to-analog
DAC	D/A converter
DCS	dynamic common source
DEM	dynamic element matching
DNL	differential nonlinearity
DR	dynamic range
DSF	dynamic source follower
DSP	digital signal processor
DT	discrete-time
DWA	data weighted averaging
ENOB	effective number of bits
EPROM	erasable programmable read-only memory
FG	foreground
FOM	figure of merit
I/O	input/output
IDAC	current-steering D/A converter
INL	integral nonlinearity
ISSCC	International Solid-State Circuits Conference
JS	junction-splitting
LMS	least-mean-square
LPF	lowpass filter
LSB	least significant bit

MASH	multi-stage noise-shaping
MDAC	multiplying DAC
MOSFET	metal-oxide-semiconductor field-effect transistor
MSB	most significant bit
NRZ	non-return to zero
NTF	noise transfer function
opamp	operational amplifier
OSR	oversampling ratio
RAMP	ring amplifier
RDAC	D/A converter using resistors
ROM	read only memory
RTZ	return-to-zero
S/H	sample-and-hold
SAR	successive approximation register
SDCT	continuous-time sigma-delta modulator
SDSC	sigma-delta modulator using switched-capacitor circuits
SFDR	spurious-free dynamic range
SMASH	sturdy MASH
SNDR	signal-to-noise-and distortion ratio
SNR	signal-to-noise ratio
SQNR	signal-to-quantization-noise ratio
STF	signal transfer function
T/H	track-and-hold
TDC	time-to-digital converter
VCO	voltage-controlled oscillator
VLSI	VLSI Circuit Symposium

1

Introduction

Decades of enormous effort put in by researchers and engineers have enabled remarkable improvements in digital LSI performance. As a result, most of the signal processing conventionally done in the analog domain is now carried out in the digital domain. However, signals in the real world are analog, which makes analog-to-digital and digital-to-analog converters, or data converters, indispensable. In this introductory chapter, the background behind the data converters is first explained, followed by a brief description of their functions, recent technological trends, and the purpose of this book. Presenting such an outline before going into details is to help readers understand smoothly about the critical points of explanations in the following chapters.

1.1 Background

A block diagram of a typical signal processing system used today is shown in Figure 1.1. The A/D converter (ADC) converts an analog input signal into a digital code, which is applied to the digital signal processor (DSP). The D/A converter (DAC) following the DSP converts the digital code into another analog signal. Taking a smartphone as an example, the ADC converts an audio signal into a digital code. The DSP generates a series of digital codes, which is then converted by the DAC into a high-frequency analog signal so that radio waves can transmit the information to a base station, which is the entrance of the communication system.

The dramatic improvement in digital signal processor performance has replaced conventional analog signal processing with digital signal processing. Another reason, which must be remembered, to make it possible is an impressive improvement in data converter performance. Data converters have a long history going back to the 16th century when an A/D conversion algorithm was suggested [1]. It is in the 1960s that data converters based on the IC technology were developed [2]. In the early days of development,

1

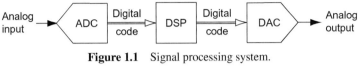

Figure 1.1   Signal processing system.

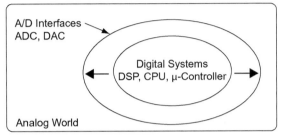

Figure 1.2   Expansion of analog/digital interface.

a data converter was made up of multiple chips, but with miniaturization of transistors, it became possible to integrate onto one chip, and high-performance conversion can be realized at a low price. More recently, it is almost unnoticeable that data converters are embedded in microcontroller chips as interfaces between analog and digital. Like many digital LSIs, Moore's law has had a significant influence on data converters.

Even though the digital signal processing becomes dominant, the real world is still analog, and data converters are indispensable on the boundary between digital and analog. Geometrically speaking, as shown in Figure 1.2, the surface area of the boundary increases as the volume of digital signal processing increases [3]. This implies that the application area of the data converter increases as the expansion of the digital domain. For example, in sensor networks and the internet of things (IoT), high-performance A/D and D/A converters are required as well as high-speed, low-power digital processors, and its role is expected to become increasingly important in the future.

1.2 Functions of Data Converters

In this section, the essential functions of the A/D converter and the D/A converter will be briefly described.

1.2.1 A/D Converter

1.2.1.1 Quantization, sampling and coding

Let us consider an example shown in Figure 1.3 to illustrate the function of the A/D converter. By using a ruler with 3-bit codes from 000 to 111, the length of object A is measured to be 110 because its right edge is in the area assigned as 110. Similarly, if the voltage V_{in} is measured with an N-bit code to obtain the digital output of $D_1 D_2 \cdots D_N$, the following equation holds

$$V_{ref} \left(D_1 2^{-1} + D_2 2^{-2} + \cdots + D_N 2^{-N} \right) = V_{in} + V_Q. \qquad (1.1)$$

Here, V_{ref}, called a reference voltage, represents the full-scale value of the input voltage. Also, in this case, N is called a bit resolution, or simply resolution. V_Q is a roundoff error and called a quantization error because this process is called quantization. Thus, quantization is the procedure to determine the section to which an analog value in question belongs. In other words, quantization is to determine D_i's in Equation (1.1).

Discretization like quantization takes place not only in the voltage domain but also in the time domain, which is called sampling. Usually, the voltage is sampled at regular intervals[1]. Only those sampled values have a meaning in digital systems. Another essential function of the A/D converter is coding, which is to assign a digital code to each segment. In this example shown in Figure 1.3, the binary code is used for coding, but other codes, such as the Grey code and the signed binary code, can also be used.

Thus, the essential functions of the A/D conversion are sampling, quantization, and coding. Compared with the sampling, the quantization is a complicated process and requires a long time. Therefore, sampling and quantization are usually carried out in this order.

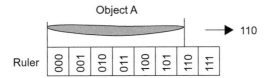

Figure 1.3 Illustration of analog-to-digital conversion.

---

[1]Non-uniformly sampled signals were discussed in [4]. Also, non-uniform sampling is sometimes attempted for rarely-occurring or occasional events.

1.2.1.2 Resolution and quantization error

The voltage corresponding to the least significant bit (LSB), V_{LSB}, is defined as

$$V_{\text{LSB}} = \frac{V_{\text{ref}}}{2^N} = V_{\text{ref}} \times 1\,\text{LSB}. \tag{1.2}$$

Here, $1\,\text{LSB} = 1/2^N$. The quantization error, V_Q, is related to V_{LSB} as

$$-\frac{1}{2}V_{\text{LSB}} < \epsilon_Q \leq \frac{1}{2}V_{\text{LSB}}. \tag{1.3}$$

Figure 1.4 shows the input/output (I/O) characteristics and quantization error of a 3-bit A/D converter. For simplicity, $V_{\text{ref}} = 1$ is assumed. Notice that the quantization error V_Q defined by Equation (1.1) is within the range $\pm 0.5\,\text{LSB}$.

   The output waveform and quantization error of a 3-bit A/D converter for a sinusoidal input are shown in Figure 1.5. The horizontal axis represents the time normalized by a sampling period of T_{s}. The reciprocal $1/T_{\text{s}}$ is

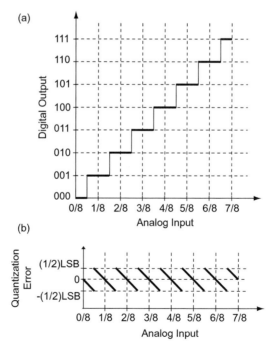

Figure 1.4   (a) Input/output characteristics and (b) quantization error of a 3-bit A/D converter for a ramp wave.

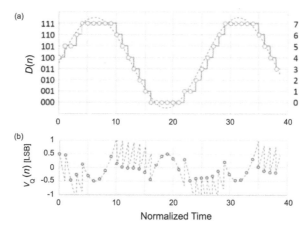

Figure 1.5   (a) Input/output characteristics and (b) quantization error of a 3-bit A/D converter for a sinusoidal-wave input.

called the sampling frequency, or the sampling rate, which is used with a unit of samples/s or S/s. In Figure 1.5(b), it is confirmed that the quantization error is within ±0.5 LSB. The dotted line, which represents the difference between the A/D converter output and the input, partially exceeds the range of ±0.5 LSB. In the A/D conversion, however, only the sampled values (denoted by open circles) have a meaning, and Equation (1.3) is satisfied.

The resolution expressed in bits is one of the most important performance specifications of the data converter. For example, to represent 1 kg within 1-g accuracy, a resolution of 3 digits in decimal is necessary. Since $2^{10} = 1024$, a resolution of 10 bits is required in this case. When measuring 1 kg within 1-mg accuracy, a resolution of 20 bits is needed. Along the vertical axis of a liquid crystal display often used in oscilloscopes, there are 256 pixels, which means that an 8-bit resolution is required to display analog input waveforms.

1.2.1.3 Circuit example

An example of a simple 3-bit A/D converter is shown in Figure 1.6. The sampling is performed by using a sample-and-hold (S/H) circuit with V_{clk} as a trigger. The input continuous-time signal $V_{in}(t)$ is discretized in the time domain into the discrete-time signal $V_{in}(n)$. The quantization of $V_{in}(n)$ is then carried out by a series of comparators, which compare the sampled voltage with threshold voltages of V_1 to V_7 to decide which segment $V_{in}(n)$ can be classified. Then, the result is encoded to obtain a 3-bit binary code.

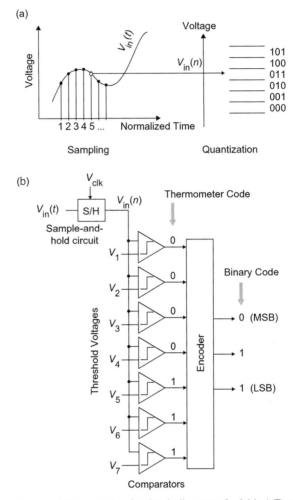

Figure 1.6   (a) Functions and (b) example of a circuit diagram of a 3-bit A/D converter.

The output of comparators is either 0 or 1, depending on the comparison result. It is assumed in this example that the threshold voltages decrease in the order from V_1 to V_7 and that the comparator output is 0 (or 1) if $V_{in}(n)$ is larger (or smaller) than the threshold voltage. The threshold voltages can be obtained, for example, by a voltage divider consisting of resistors connected in series, as will be shown in Section 5.3. In this example,

$$V_4 > V_{in}(n) > V_5 \tag{1.4}$$

is assumed. Therefore, the output of the three comparators at the bottom is 1, and the output of the other four comparators is 0. When these values are aligned from top to bottom, the output code obtained is 0000111. Such a code is called a thermometer code because it resembles the behavior of mercury thermometer; as the input increases, the boundary between 0 and 1 moves upward, which looks like a mercury thermometer when the temperature rises. The thermometer code is a redundant representation, and usually, it is encoded into a compact one, such as the binary code as shown in Table 1.1, to obtain the final digital output.

Figure 1.7 is a circuit symbol of the A/D converter. Along with the analog input value V_{in}, the reference voltage V_{ref}, and the clock signal V_{clk} are used for sampling.

1.2.2 D/A Converter

1.2.2.1 Input/output characteristics

An example of D/A conversion is illustrated in Figure 1.8. This shows a way in which a 3-bit input code is converted into an analog quantity by using a set of bars with binary-weighted lengths. Each bar corresponds to each bit of the input code. By joining together the bars corresponding to an input of 1, the analog output is obtained.

Table 1.1 Binary code b_i and thermometer code d_i

$b_1 b_2 b_3$	$d_1 d_2 d_3 d_4 d_5 d_6 d_7$
000	0000000
001	0000001
010	0000011
011	0000111
100	0001111
101	0011111
110	0111111
111	1111111

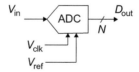

Figure 1.7 Circuit symbol of A/D converter.

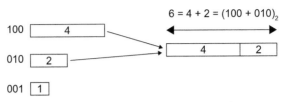

Figure 1.8   Illustration of digital-to-analog conversion.

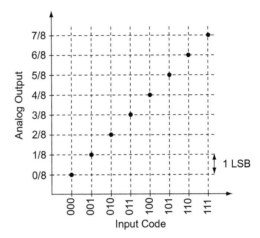

Figure 1.9   Input/output characteristics of a 3-bit D/A converter.

If the input digital code D_{in} is $D_1 D_2 \cdots D_N$, the analog output value, V_{out}, is represented as

$$V_{\mathrm{out}} = V_{\mathrm{ref}} \left(D_1 2^{-1} + D_2 2^{-2} + \cdots + D_N 2^{-N} \right). \tag{1.5}$$

Unlike the A/D converter, the quantization error does not appear in the equation. The input/output characteristics of a 3-bit D/A converter are shown in Figure 1.9.

1.2.2.2 Circuit example

A circuit example of a 3-bit D/A converter is shown in Figure 1.10. Candidates of the analog output from 0 V to $(7/8)\, V_{\mathrm{ref}}$ with a $(1/8)\, V_{\mathrm{ref}}$ step are prepared by dividing V_{ref} with a resistor ladder. One of them is selected by the MOSFET switches according to the digital input code so that the node corresponding to the code is connected to the output. For example, if the input is 110, b_1, b_2, and $\bar{b}_3$ are HIGH, so that $V_x (= (3/4) V_{\mathrm{ref}})$ is connected to V_{out}. The circuit symbol of the D/A converter is shown in Figure 1.11.

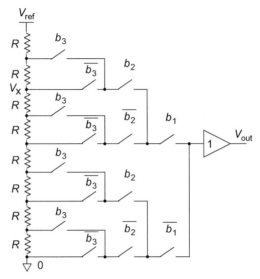

Figure 1.10 Circuit example of a 3-bit D/A converter.

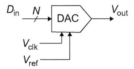

Figure 1.11 Circuit symbol of D/A converter.

1.2.3 D/A Converter used in A/D Conversion

While D/A converters operate by itself, A/D converters often operate with the assistance of a D/A converter. For example, consider a procedure to measure the weight of an object B by using binary-weighted weights. It is assumed here that the weight of the object is less than 8. First, the object weight is compared with the weight of 4 as shown in Figure 1.12. If the object is heavier than 4, the weight of 2 is added. If the object is heavier than $4+2$, the weight of 1 is added. In this example, the sum of the weights becomes heavier than the object. Then it is concluded that the weight of the object is between 6 and 7, and the digital output is obtained as 110. This means that during the A/D conversion, a D/A converter consisting the weights is used. The input of the D/A converter is 100 as the first guess, followed by 110 and 111. This is called a binary search algorithm, and details of circuits implementing this method will be described in Section 5.4.

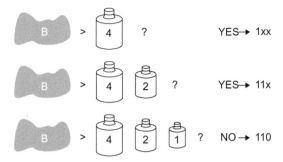

Figure 1.12   Procedure to weigh using weights.

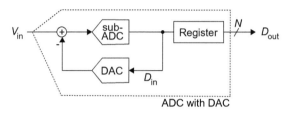

Figure 1.13   Internal D/A converter used for A/D conversion.

Figure 1.13 is a block diagram for performing the A/D conversion procedure described above. The digital output D_{out} is determined in order from the most significant bit (MSB) to the least significant bit (LSB). During the comparison cycle, the output of the D/A converter gradually approaches V_{in}. The input to the D/A converter after the final comparison is the output of the A/D converter. In practice, the register is used to store the estimated values and to generate the final output code.

1.3 Trends

1.3.1 Technology and Architectures

The technology used for fabricating A/D converters is shown in Figure 1.14. Data were taken from papers published at the International Solid-State Circuits Conference (ISSCC) and the VLSI Circuit Symposium (VLSI), both of which are international conferences that are well-known for many excellent papers on integrated circuits[2]. This figure shows that the state-of-the-art technology nodes developed for digital LSIs were introduced to A/D converters

---

[2]A list of data on A/D converters reported at ISSCC and VLSI is available on the website [5]. The author drew the figures in this section based on it.

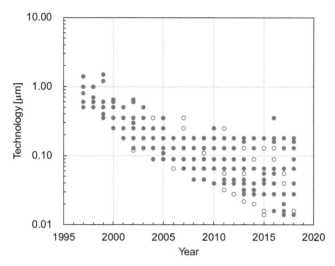

Figure 1.14 Technology used for implementing A/D converters published in ISSCC (closed circuits) and VLSI Symposium (open circuits).

almost at the same time. Nevertheless, it should be noted that technologies of generations ago from the viewpoints of digital LSIs, such as a 0.18 μm technology, are still alive. This is because the required performance is diversifying with the expansion of application fields. On the one hand, the priority is to increase the speed by making full use of the scaled-down CMOS technology, and on the other hand, it is to lower the cost by using a matured and inexpensive process while keeping the necessary performance. It can be said that bipolarization is progressing.

Figure 1.15 shows the number of papers on various architectures of A/D converters. The number of published papers has increased since 2000, indicating high activities of research and development in this field. In particular, from around 2006, the number of papers on successive approximation (SAR) increased. This old architecture is attracting the limelight now because of its superior low power consumption. Regarding the ΔΣ A/D converters, the focus is not only on the conventional discrete-time type (SDSC) but also on the continuous-time type (SDCT). The latter is suitable for high-speed operation aiming at communication applications. Details of each architecture will be explained in the following chapters, but here, it should be mentioned that intensive research and development are conducted on several architectures as well as hybrid ones combining them.

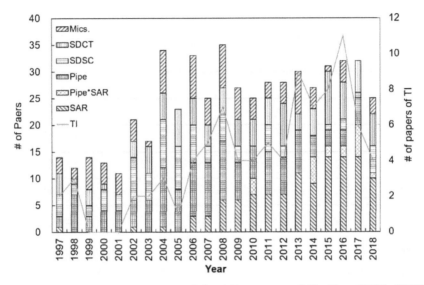

Figure 1.15 Trend of architectures used for A/D converter. SAR, Pipe, SDSC, SDCT, TI, SAR*Pipe, and Mics. stand for successive approximation, pipelined, discrete-time $\Delta\Sigma$, continuous-time $\Delta\Sigma$, time-interleaved, pipelined successive approximation, and others, respectively.

Furthermore, note that architectures shown in Figure 1.15 attract attention at the forefront of research and development. Many A/D converters used in commercial equipment are based on conventional architectures, such as integration (dual-slope) type and flash type. Therefore, in this book, these architectures will also be explained.

1.3.2 Performance and Applications

Today, since A/D converters are applied in various manners, performance specifications to be satisfied are also diverse. Among them, the most basic specifications are conversion speed, resolution, and power consumption. The conversion speed is related to the input signal bandwidth. According to the Nyquist sampling theorem described in Section 2.1, if the conversion speed is high, the input bandwidth expands, and A/D conversion for high-frequency signals becomes possible. The resolution is expressed by the number of bits or the signal-to-noise ratio (SNR) where the quantization error is considered as noise. The power consumption depends on the operation speed and resolution. For comparing the performance, a figure of merit

(FOM) is proposed as follows:

$$\text{FOM} = \frac{P}{f_s \cdot 2^N}.$$ (1.6)

Here, P, f_s, N are the power, the sampling frequency, and the bit resolution, respectively. It is inferred from the analogy with the CMOS logic circuit that the power consumption increases in proportion to the operating frequency. It can also be expected that the power consumption will increase if the resolution increases. Therefore, it is understood that Equation (1.6) is reasonable as the normalized power consumption. The detailed discussion of FOM will be described in Section 7.1.

The power consumption of A/D converters is shown in Figure 1.16 as a function of the bit resolution. The vertical axis is the power consumption divided by the sampling frequency under the Nyquist condition, f_{snyq}, which means the energy required per conversion. The horizontal axis represents the resolution called an effective number of bits (ENOB), which is described in detail in Section 2.2. As shown by the arrow, the lower right area of the plot is the target area: high bit resolution with low energy consumption. As shown in this figure, based on numerous experimental data, more energy is required to increase the resolution. The asymptote qualitatively agrees with the expression of FOM (Equation (1.6)).

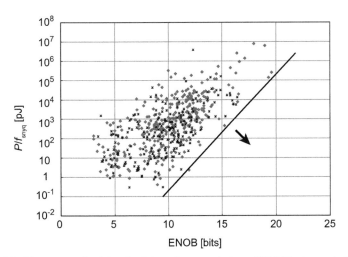

Figure 1.16 Energy required for obtaining given resolutions (ENOB). Arrow indicates the direction to go. Asymptote is explained in Section 7.1.

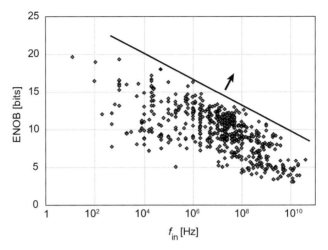

Figure 1.17   Relation between the resolution (ENOB) and maximum input frequency. Arrow indicates the direction to go. Asymptote is explained in Section 7.1.

The relationship between the bit resolution and the maximum input frequency f_{in} is shown in Figure 1.17. The vertical axis represents the effective number of bits (ENOB). The arrow directing the upper right indicates the target area: high resolution with high speed. However, the ENOB evaluated for actual circuits decrease as the operation speed increases, as indicated by the ascending line to the lower right. When examining the published date, the asymptote moves to the upper right direction year by year, but the moving speed is rather slow compared with that in Figure 1.16. Jitter or fluctuation in the sampling time is responsible for the decrease in the resolution, which will be explained in Sections 2.1 and 3.1.

The application field of A/D converters extends over a wide range from the high-precision instrumentation application to ultra-high-speed optical communication application, as shown in Figure 1.18. It can be seen that a wide range of application fields are lined up along the asymptote shown in Figure 1.17.

1.4 Purpose of this Book

The primary purpose of this book is to understand the fundamentals of data conversion as well as to develop the ability to keep up with future progress in this exciting field. The topics are selected by considering the

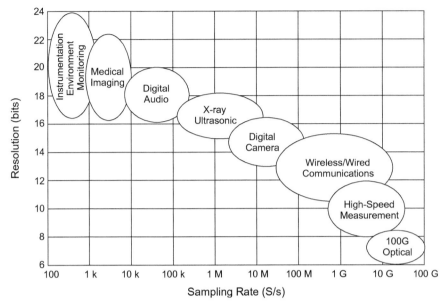

Figure 1.18 Application fields of A/D converters (after [6]).

recent technology trends mentioned above. First, signal processing and basic circuit blocks will be explained. Next, various conversion algorithms and their circuit implementations will be described. This is the main part of this book. Finally, low-power amplifiers, hybrid configurations, and digitally-assisted techniques will be touched on to introduce the latest trends.

Those who intend to engage in research and development of data converters are supposed as readers of this book. Also, this book will be helpful for engineers who intend to develop new equipment or systems by exploiting data converters. For example, it is necessary to gather a wide range of experts from system designers to device engineers for building innovative sensor networks and the IoT. Understanding the basics described here makes interactive communication smooth, which will result in achieving optimized system development.

When writing this book, I paid special attention to the following points:

1. To explain basic items intensively and systematically.
2. To include conversion algorithms and circuit design techniques reflecting scaled-down CMOS technology.
3. To make a bridge between the basics and the cutting edge by citing many references.

It is desirable for readers to be familiar with basic analog IC design, including operational amplifiers (opamps), and digital signal processing using z-transform.

As for data converters, many excellent books [7–12], review articles [13–20], and a handbook [21] are already published. Books on oversampling A/D converters [22–26] are available. Textbooks on analog integrated circuit design [27–30] are also highly recommended. Furthermore, topics of data converters are discussed in various articles and meetings[3]. I would like to encourage readers to refer to them for further understanding.

---

[3]For example, IEEE Solid-State Circuits Magazine Vol. 7, No. 3 No. 3, IEEE International Solid-State Circuits Conference (ISSCC) 2016 and 2017 forum, 2006 SC1 to SC4 (A/D converter performance limit, pipeline, $\Delta\Sigma$ modulator, low voltage), 2007 T2 (Continuous time $\Delta\Sigma$ modulator), 2008 T2 and 2014 T5 (Pipelined), 2009 T6 (SAR), 2012 SC3 (A/D converter performance limit), 2015 SC20 (Miniaturized CMOS A/D converter), 2015 T5 (High-speed SAR) (SC and T are for short courses and tutorials. Refer to the text for abbreviations).

2

Basic Principles

In this section, sampling and quantization, which are two of the most basic functions of data converters, are described. Regarding the sampling that discretizes a continuous-time analog signal in the time domain, the following topics are covered:

1. For the original analog signal to be correctly reproduced from the sampled data, the sampling frequency must be at least twice the signal bandwidth (the sampling theorem).
2. After sampling, signal components appear at different frequencies from the original one (the aliasing). Two signals that have overlapped in the frequency domain cannot be distinguished from each other. Therefore, an anti-aliasing filter should be used before sampling.
3. Oversampling and undersampling have unique features that are useful in specific applications.
4. Sampling-time jitter severely deteriorates the signal-to-noise ratio (SNR).

Quantization is an approximation of a continuous analog quantity to one of the predetermined discrete values depending on its amplitude. Following topics are described:

1. Quantization errors limit the SNR of data converters.
2. For a sinusoidal input signal, the relation between the bit resolution N and the SNR is written as

$$\text{SNR (in dB)} = 6.02N + 1.76. \tag{2.1}$$

These are premises for understanding the following chapters. Readers who are already familiar with these topics can skip this chapter.

2.1 Sampling

Sampling is to pick up instantaneous values of continuous-time analog signals one by one at predetermined time intervals. Figure 2.1 shows an example of sampling the analog signal $f(t)$. The discrete-time signal, $f^*(t)$, is obtained by sampling with a time interval of T_s. T_s is called a sampling period. $f^*(t)$ is defined only at $t = nT_\mathrm{s}$, and is often denoted as $f^*(nT_\mathrm{s})$ or $f^*(n)$, where n is an integer. Both of them are equal to $f(nT_\mathrm{s})$.

2.1.1 Sampling Theorem and Aliasing

First, let us consider to sample three sinusoidal signals with different frequencies by using the same sampling frequency. Figures 2.2(a)-(c) show results for 14-, 36-, and 64-Hz signals sampled with a sampling frequency of 50 Hz. It turns out that the sampled values are all the same as shown in Figure 2.2(d). This indicates that it is not possible to uniquely reproduce the original signal from the sampled values. Therefore, it is worth considering changes in spectra before and after the sampling for more general cases.

2.1.1.1 Spectra before and after sampling

In Figure 2.3, $f(t)$ represents an analog signal continuously changing with time. The waveform of the signal after the sampling with a period of T is also shown as $f^*(t)$ by assuming that each sampled value is held for τ. $f^*(t)$ consists of a series of pulses, in which the n-th pulse is represented as

$$f_n(t) = kf(nT)[u(t - nT) - u(t - nT - \tau)]. \qquad (2.2)$$

Here, $u(t - nT)$ is the step function defined as

$$u(t - nT) = \begin{cases} 0 & (t < nT) \\ 1 & (t \geq nT) \end{cases}. \qquad (2.3)$$

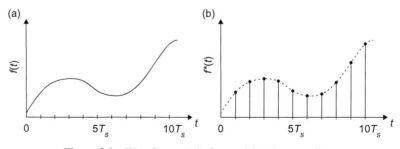

Figure 2.1   Waveforms (a) before and (b) after sampling.

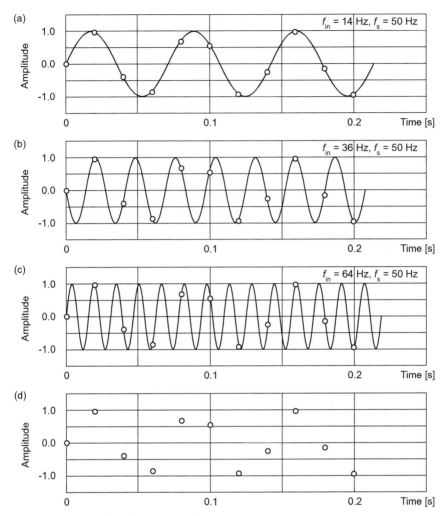

Figure 2.2 Sampling of three sinusoidal inputs having different frequencies.

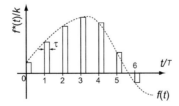

Figure 2.3 Sampling of an analog signal (t).

k is a normalization constant equal to $k = 1/\tau$ so that the pulse area is kept constant.

The series of sampled pulses $f^*(t)$ is expressed as

$$f^*(t) = \sum_{n=-\infty}^{\infty} f_n(t) \tag{2.4}$$

$$= k \sum_{n=-\infty}^{\infty} f(nT)\left[u(t - nT) - u(t - nT - \tau)\right]. \tag{2.5}$$

The Laplace transform of $f^*(t)$ is then

$$F^*(s) = k \sum_{n=-\infty}^{\infty} f(nT)\left[\frac{1}{s}e^{-nTs} - \frac{1}{s}e^{-(nT+\tau)s}\right]$$

$$= k\frac{1 - e^{-\tau s}}{s} \sum_{n=-\infty}^{\infty} f(nT)e^{-nTs}. \tag{2.6}$$

When the pulse width is infinitely narrowed ($\tau \to 0$),

$$F^*(s) = k\frac{1 - (1 - \tau s)}{s} \sum_{n=-\infty}^{\infty} f(nT)e^{-nTs}$$

$$= \sum_{n=-\infty}^{\infty} f(nT)e^{-nTs} \tag{2.7}$$

is obtained. This equation indicates that $F^*(s)$ is determined only by the sampled values, $f(nT)$, and does not depend on the signal values between the sampling points. The spectrum of $f^*(t)$ for the extremely narrowed pulse width can be obtained by replacing s in Equation (2.7) with $j\omega$ as follows:

$$F^*(j\omega) = \sum_{n=-\infty}^{\infty} f(nT)e^{-j\omega nT}. \tag{2.8}$$

Alternatively, Equation (2.8) can be derived using the Fourier transform as follows. The sampling pulse sequence $f^*(t)$ for the extremely narrowed pulse width is

$$f^*(t) = \sum_{n=-\infty}^{\infty} f(t)\delta(t - nT), \tag{2.9}$$

and its Fourier transform is written as

$$\int_{-\infty}^{\infty} f^*(t)e^{-j\omega t}dt = \int_{-\infty}^{\infty}\left(\sum_{n=-\infty}^{\infty} f(t)\delta(t-nT)\right)e^{-j\omega t}dt$$

$$= \sum_{n=-\infty}^{\infty} f(nT)e^{-j\omega nT}, \tag{2.10}$$

which is equal to Equation (2.8). By using the Fourier transform formula of the delta function, Equation (2.10) can be rewritten as

$$F^*(j\omega) = \frac{1}{T}\sum_{k=-\infty}^{\infty} F\left(j\omega - jk\frac{2\pi}{T}\right). \tag{2.11}$$

Here, $F(j\omega)$ on the right side represents the Fourier transform of the input continuous signal

$$F(j\omega) \equiv \int_{-\infty}^{\infty} f(t)e^{-j\omega t}dt. \tag{2.12}$$

Equation (2.11) shows that after sampling, the spectrum before sampling is repeated with a period of the angular frequency $\frac{2\pi}{T}$. In other word, the spectrum after sampling is a repetition of the original spectrum with a period of the sampling frequency f_s. The details in deriving Equation (2.11) will be described in Subsection 2.1.1.5.

In the example shown at the beginning of this section, since the 14-Hz sine wave has a spectrum with peaks at ± 14 Hz, the spectrum shown in Figure 2.4 is obtained after the sampling[1]. New peaks appear at 36 Hz, 64 Hz, and so on,

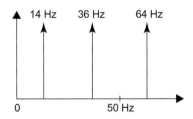

Figure 2.4 Spectrum of a sinusoidal input with an input frequency of 14 Hz and a sampling frequency of 50 Hz.

[1]The magnitude of the Fourier transform of a real function is symmetric with respect to the imaginary axis.

besides 14 Hz. Since a similar repetition in the frequency domain takes place for 36-Hz and 64-Hz sine waves, their spectra coincide with one another after the sampling with a frequency of 50 Hz so that they cannot be distinguished.

2.1.1.2 Reproduction of original signal

To find a way to reproduce the original continuous-time signal from the sampled values, let us consider the main lobe, which is a part of the spectrum within a range of $[-\pi/T, \pi/T]$ and defined as

$$\hat{F}(j\omega) \equiv H(j\omega) F^*(j\omega). \tag{2.13}$$

Here, the transfer function $H(j\omega)$ is represented as

$$H(j\omega) \equiv \begin{cases} T & (|\omega| \leq (\pi/T)) \\ 0 & (|\omega| > (\pi/T)) \end{cases}. \tag{2.14}$$

If the spectrum of the continuous signal before sampling does not spread out of the interval $[-\pi/T, \pi/T]$ like $F_A(j\omega)$ in Figure 2.5, $\hat{F}(j\omega)$ of Equation (2.13) is equal to the spectrum $F_A(j\omega)$ of the continuous signal before sampling, so $\hat{f}(t)$ should also be equal to the original continuous signal $f(t)$. Thus,

$$f(t) = \hat{f}(t) = \int_{-\infty}^{\infty} \hat{F}(j\omega)e^{j\omega t}dt. \tag{2.15}$$

On the other hand, if the spectrum of the original signal spreads outside the interval $[-\pi/T, \pi/T]$ as shown by $F_B(j\omega)$ in Figure 2.5, the components that go beyond the interval overlap each other. The spectrum after sampling

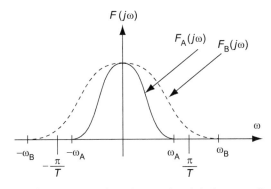

Figure 2.5   Spectra of continuous signals before sampling.

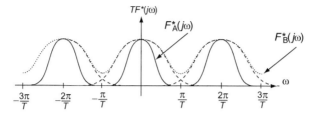

Figure 2.6 Spectra obtained after sampling two input signals shown in Figure 2.5.

is shown as $F_B^*(j\omega)$ in Figure 2.6. Therefore, even if the signal spectrum is multiply by the transfer function $H(j\omega)$, it is different from the original continuous signal spectrum $F_B(j\omega)$. Therefore, the inverse Fourier transform $\hat{f}(t)$ is also different from the original continuous signal $f(t)$.

Summarizing the explanation mentioned above, it is necessary for the spectrum of the original continuous signal to be within the interval $[-\pi/T, \pi/T]$ to reproduce the original continuous signal from the sampled discrete values. If Equation (2.15) is calculated under this condition, the following expression can be obtained

$$f(t) = \sum_{n=-\infty}^{\infty} f(nT)\frac{\sin\left[(\pi/T)(t-nT)\right]}{(\pi/T)(t-nT)}. \tag{2.16}$$

Here, $f(nT)$ is the sampled discrete value, and $f(t)$ is the original continuous signal. Derivation details will be shown in Subsection 2.1.1.5.

To explain with the example mentioned at the beginning of this section, as shown by Figure 2.7, if only the main lobe within a dashed frame is extracted, then, the original 14 Hz input sine wave can be obtained after performing the inverse Fourier transform.

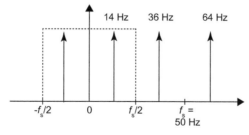

Figure 2.7 Spectrum of a sampled sinusoidal signal. Peaks in the dotted frame correspond to the main lobe.

2.1.1.3 Sampling theorem

As is mentioned in the previous subsection, the condition for the input continuous signal to be reproduced from the sampling values is that the spectrum of the original signal is within the interval $[-\pi/T, \pi/T]$. In other words, the maximum value of the frequency of input signal components, or the signal bandwidth f_B, should be less than half the sampling frequency: $f_s \geq 2f_B$, where f_s is the sampling frequency. This is called the Nyquist sampling theorem, or the sampling theorem in short. If $f_s = 2f_B$, f_s is called the Nyquist rate. When the sampling frequency is f_s, the frequency $f_s/2$ is called the Nyquist frequency. If the signal bandwidth is less than the Nyquist frequency, the original signal can be reproduced from the sampled values. The Nyquist rate and the Nyquist frequency are sometimes quite misleading terms, but they are always used in this manner. Also, when $f_s = 2f_B$, it is said that the measurement is performed under the Nyquist condition. In practice, frequencies slightly above $2f_B$ are often used for Nyquist-condition measurements.

2.1.1.4 Aliasing and anti-aliasing filter

Consider what happens when sampling the signal F_c under the presence of a sine wave of frequency f_1 as shown in Figure 2.8(a). According to Equation (2.11), after sampling at a frequency of f_s, the f_1 sine wave is periodically repeated with a period of f_s, as shown in Figure 2.8(b). The different-frequency signal is called an "alias" of the original signal, and "aliasing" means that the signal appears at frequencies different from the

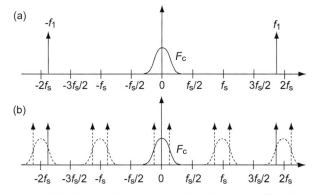

Figure 2.8　Spectra (a) before and (b) after sampling of a signal (F_c) with an out-of-band interference one (f_1). Note aliasing of the interferer (f_1) into the signal band.

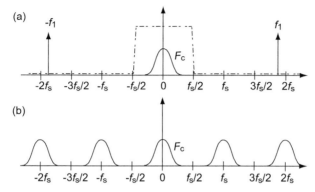

Figure 2.9 Spectra (a) before and (b) after sampling using an anti-aliasing filter.

original one. Of course, the continuous spectra (F_c) is also repeated with a period of f_s and spread out.

If the out-of-band signal components, like f_1 in Figure 2.8, overlap an input signal due to the aliasing, the spectrum of the original signal is deformed, and cannot be correctly reproduced. To prevent this from happening, it is necessary to use a filter having the characteristics indicated by the dot-dashed line of Figure 2.9(a). This filter removes signals outside the signal band before sampling. Then no additional signals are folded back to the original signal as shown in Figure 2.9(b), and a correct signal reproduction becomes possible. Such a filter is called an anti-aliasing filter. After sampling, F_c expands periodically, but as indicated by Equation (2.13), the inverse Fourier transform shown in Equation (2.16) for only the interval of $\pm f_s/2$, can reproduce the original signal properly.

2.1.1.5 Derivation of equations
The following is the derivation of equations used above. Readers who do not need this can skip to the next section.

2.1.1.5.1 *Derivation of Equation (2.11)*
By expressing the sampled values as

$$f^*\left(t\right) = \sum_{n=-\infty}^{\infty} f\left(t\right)\delta\left(t-nT\right) \equiv f\left(t\right)s(t), \qquad (2.17)$$

let us consider the Fourier transforms of $f(t)$ and $s(t)$. Based on the Fourier transform of the delta function[2],

$$F(j\omega) = \int_{-\infty}^{\infty} f(t)e^{-j\omega t}dt \tag{2.18}$$

and

$$S(j\omega) = \int_{-\infty}^{\infty} s(t)e^{-j\omega t}dt = \frac{2\pi}{T} \sum_{k=-\infty}^{\infty} \delta\left(\omega - k\frac{2\pi}{T}\right) \tag{2.19}$$

are obtained. Carrying out the Fourier transform of the product of these functions results in the following equation:

$$
\begin{aligned}
F^*(j\omega) &\equiv \int_{-\infty}^{\infty} f(t)\, s(t)e^{-j\omega t}dt \\
&= \int_{-\infty}^{\infty} f(t)\left[\frac{1}{2\pi}\int_{-\infty}^{\infty} S(j\omega')e^{j\omega' t}d\omega'\right]e^{-j\omega t}dt \\
&= \int_{-\infty}^{\infty} f(t)\left[\frac{1}{2\pi}\int_{-\infty}^{\infty} \frac{2\pi}{T}\sum_{k=-\infty}^{\infty}\delta\left(\omega' - k\frac{2\pi}{T}\right)e^{j\omega' t}d\omega'\right]e^{-j\omega t}dt \\
&= \frac{1}{T}\int_{-\infty}^{\infty}\left[\int_{-\infty}^{\infty} f(t)\sum_{k=-\infty}^{\infty}\delta\left(\omega' - k\frac{2\pi}{T}\right)e^{j\omega' t}e^{-j\omega t}dt\right]d\omega' \\
&= \frac{1}{T}\int_{-\infty}^{\infty}\sum_{k=-\infty}^{\infty}\delta\left(\omega' - k\frac{2\pi}{T}\right)\left[\int_{-\infty}^{\infty} f(t)\,e^{j\omega' t}e^{-j\omega t}dt\right]d\omega' \\
&= \frac{1}{T}\int_{-\infty}^{\infty}\sum_{k=-\infty}^{\infty}\delta\left(\omega' - k\frac{2\pi}{T}\right)F(j\omega - j\omega')d\omega' \\
&= \frac{1}{T}\sum_{k=-\infty}^{\infty}F(j\omega - jk\frac{2\pi}{T}).
\end{aligned}
\tag{2.20}
$$

2.1.1.5.2 *Fourier transform of the delta function*

Consider the Fourier transform of a series of delta functions represented as

$$s(t) = \sum_{n=-\infty}^{\infty} \delta(t - nT). \tag{2.21}$$

---

[2]This will be explained shortly.

Since $s(t)$ is a periodic function with the sampling period of T, $s(t)$ can be represented by a Fourier series as follows:

$$s(t) = \sum_{k=-\infty}^{\infty} c_k e^{jk\frac{2\pi t}{T}}. \tag{2.22}$$

The Fourier coefficients c_k is calculated as

$$
\begin{aligned}
c_k &= \frac{1}{T} \int_{-T/2}^{T/2} s(t) e^{-jk\frac{2\pi t}{T}} dt \\
&= \frac{1}{T} \int_{-T/2}^{T/2} \sum_{n=-\infty}^{\infty} \delta(t - nT) e^{-jk\frac{2\pi t}{T}} dt = \frac{1}{T},
\end{aligned} \tag{2.23}
$$

and then $s(t)$ is

$$s(t) = \frac{1}{T} \sum_{k=-\infty}^{\infty} e^{jk\frac{2\pi t}{T}}. \tag{2.24}$$

Therefore, the Fourier transform is written as

$$
\begin{aligned}
S(j\omega) &= \int_{-\infty}^{\infty} \frac{1}{T} \sum_{k=-\infty}^{\infty} e^{jk\frac{2\pi t}{T}} e^{-j\omega t} dt \\
&= \frac{1}{T} \int_{-\infty}^{\infty} \sum_{k=-\infty}^{\infty} e^{jk\frac{2\pi t}{T} - j\omega t} dt \\
&= \frac{1}{T} \sum_{k=-\infty}^{\infty} \int_{-\infty}^{\infty} e^{jk\frac{2\pi t}{T} - j\omega t} dt \\
&= \frac{1}{T} \sum_{k=-\infty}^{\infty} \int_{-\infty}^{\infty} e^{j\left(k\frac{2\pi}{T} - \omega\right)t} dt \\
&= \frac{2\pi}{T} \sum_{k=-\infty}^{\infty} \delta\left(\omega - k\frac{2\pi}{T}\right).
\end{aligned} \tag{2.25}
$$

2.1.1.5.3 *Derivation of Equation (2.16)*

The main lobe of the spectrum $F^*(j\omega)$ is

$$\hat{F}(j\omega) \equiv H(j\omega)F^*(j\omega) = H(j\omega) \sum_{n=-\infty}^{\infty} f(nT)e^{-j\omega nT}, \tag{2.26}$$

where $H(j\omega)$ is the transfer function defined by Equation (2.14). Since the inverse Fourier transform of $H(j\omega)$ is written as

$$
\begin{aligned}
h(t) &= \frac{1}{2\pi} \int_{-\infty}^{\infty} H(j\omega) e^{j\omega t} d\omega \\
&= \frac{T}{2\pi} \int_{-(\pi/T)}^{(\pi/T)} e^{j\omega t} d\omega = \frac{T}{2\pi} \left[\frac{e^{j\omega t}}{jt} \right]_{-(\pi/T)}^{(\pi/T)} \\
&= \frac{T}{2\pi jt} \left[e^{j(\pi/T)} - e^{-j(\pi/T)} \right] \\
&= \frac{\sin(\pi/T)}{\pi t/T},
\end{aligned}
\tag{2.27}
$$

the inverse Fourier transform of $\hat{F}(j\omega)$ is obtained as

$$
\begin{aligned}
\hat{f}(t) &\equiv h(t) \oplus f^*(t) = \int_{-\infty}^{\infty} h(\tau) f^*(\tau - t) d\tau \\
&= \int_{-\infty}^{\infty} \frac{\sin(\pi\tau/T)}{\pi\tau/T} \sum_{n=-\infty}^{\infty} f(\tau - t) \delta(\tau - t - nT) d\tau \\
&= \sum_{n=-\infty}^{\infty} f(nT) \frac{\sin((\pi/T)(t - nT))}{(\pi/T)(t - nT)}.
\end{aligned}
\tag{2.28}
$$

2.1.2 Oversampling and Undersampling

According to the sampling theorem described in the previous subsection, the waveform of the original signal can be correctly reproduced by using a sampling frequency which is equal to or more than $2f_B$. Here, f_B is the signal bandwidth. Figure 2.10(a) indicates that sampling is performed with $f_{s1} = 2f_B$. In this case, the anti-aliasing filter, the transfer characteristics of which sharply drop to zero at $f_s = 2f_B$, is required[3]. In practice, it is difficult to design such filter characteristics. Also, such a filter frequently results in unwanted ringing. It is then usually adopted a sampling frequency of f_{s2} slightly higher than $2f_B$ as shown in Figure 2.10(b). This relaxes the requirement on the filter characteristics and reduces the burden on the filter design. Figure 2.10(c) shows the case for a frequency of f_{s3} ($> f_{s2}$), where the filter condition can be more relaxed.

---

[3]This is commonly known as brick wall characteristics.

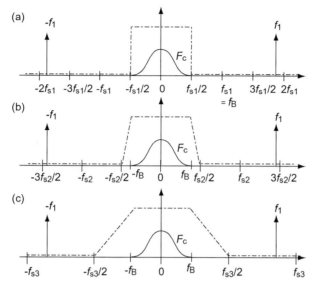

Figure 2.10   Spectra obtained with different sampling frequencies, $s1$ to $s3$. (a) represents Nyquist rate sampling, while (b) and (c) represent oversampling. $_1$ represents an interferer. Dash-dotted lines represent the characteristics of anti-aliasing filters.

In general, sampling at frequencies higher than $2f_B$ is called oversampling, and OSR $= \frac{f_s}{2f_B}$ is called the oversampling ratio. The larger the OSR, the more relaxed the constraint of the anti-aliasing characteristics. Although the details are described in Section 6.1, there is another merit that the quantization noise in the signal band can be reduced by increasing the OSR. Oversampling often implies $\Delta\Sigma$-type A/D converters, as described in Chapter 6. As explained above, in the $\Delta\Sigma$-type A/D converter, the requirement for the anti-aliasing filter characteristics is substantially relaxed.

In contrast, sampling at $f_s \leq 2f_B$ is called undersampling. The signal components are folded into the low-frequency region. Its unique feature is to shift the signal frequency down to the low-frequency region. This is sometimes called down-conversion[4]. An example of downsampling is shown in Figure 2.11. It is assumed that the original signal components are only in the frequency region $[f_s, 3f_s/2]$ and that there are no other signals.

---

[4]This is the same function usually obtained in the analog domain called mixing. Conventionally, mixing is carried out by multiplying the original signal by a sinusoidal wave generated by a local oscillator.

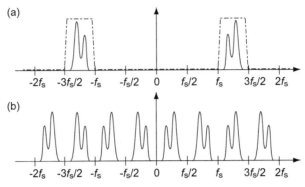

Figure 2.11   Spectra (a) before and (b) after undersampling. The dot-dashed line represents bandpass characteristics.

An anti-aliasing filter having a band-pass characteristic can avoid the overlapping with other signals. Note that the sampling circuit needs high-frequency characteristics covering the signal band before the undersampling. The undersampling technique draws attention in recent years as an interesting technique called analog-to-information conversion, which tries to extract necessary information from high-frequency analog signals directly. Interested readers should refer to the article [31] listed at the end of the book.

2.1.3 Jitter and SNR

A clock generation circuit is necessary to sample the input signal at predetermined time intervals. In the explanation in previous sections, it is assumed that the circuit generates an ideal clock signal, and the pulse interval does not fluctuate. However, since the interval of pulses generated by an actual circuit varies due to various factors, the sampling time also fluctuates. This fluctuation is called jitter[5]. An example of jitter is shown in Figure 2.12. If the sampling time changes from T to $T + \delta(T)$, there occurs an error of Δx_1 in the sampled value. Digital systems, which process sampled values, assume that they are sampled at ideal intervals without jitter. Therefore, it is necessary to evaluate the errors due to jitter. When the input signal rapidly changes, the error caused by jitter also increases. In this example, Δx_2 is the largest among others. In particular, to accurately sample a high-frequency signal, it is essential to suppress the jitter.

---

[5]The mechanism generating jitter is described in Subsection 3.1.8.

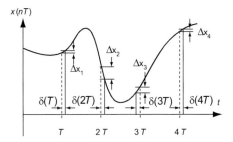

Figure 2.12 Jitter in sampling time.

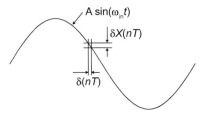

Figure 2.13 Effect of jitter on a sinusoidal input.

Consider sampling a sine wave with an input amplitude and angular frequency of A and ω_{in}, respectively. The effect of jitter is shown in Figure 2.13. If the sampling time of $t = nT$ changes by $\delta(nT)$ due to jitter, the error in the sampled value is

$$\delta X\,(nT) = A\omega_{in}\delta(nT)\cos(\omega_{in}nT). \qquad (2.29)$$

If $\delta(nT)$ is considered as the random variable $\delta_{ji}(t)$, the fluctuation of sampled values due to jitter is

$$< x_{ji}(t)^2 > \; = \; < [A\omega_{in}\cos(\omega_{in}nT)]^2 >< \delta_{ji}(t)^2 >$$

$$= \frac{A^2\omega_{in}^2}{2} < \delta_{ji}(t)^2 > . \qquad (2.30)$$

If the variation due to jitter can be dealt with as noise, the signal-to-noise ratio (SNR) is obtained as

$$\mathrm{SNR}_{ji}\ (\text{in dB}) \equiv 10\cdot\log\frac{\left(\frac{A^2}{2}\right)}{\frac{A^2\omega_{in}^2}{2}\,\delta_{ji}(t)^2}$$

$$= -10\cdot\log\left(\omega_{in}^2 < \delta_{ji}(t)^2 >\right). \qquad (2.31)$$

The SNR decreases due to jitter.

The slope is maximum when the signal crosses zero, and the influence of jitter is the largest. Therefore, the maximum value of the error is

$$\Delta x \left(nT\right)|_{\max} = A\omega_{\mathrm{in}}\delta t_{\max} = A2\pi f_{\mathrm{in}}\delta t_{\max}. \tag{2.32}$$

In data conversion, it is necessary to make this error smaller than the minimum resolution $V_{\mathrm{LSB}} = 2A/2^N$ corresponding to the least significant bit. This is represented as

$$A2\pi f_{\mathrm{in}}\delta t_{\max} < \frac{2A}{2^N}. \tag{2.33}$$

Therefore, the maximum allowable jitter $\delta t_{\max}$ required to obtain N-bit resolution is

$$\delta t_{\max} < \frac{1}{2^N \pi f_{\mathrm{in}}}. \tag{2.34}$$

The higher the frequency of the input wave, the more the SNR deteriorates as shown in Figure 2.14. A typical value of jitter is around 1 ps[6]. Thus, for sampling at 100 MS/s, or for a 50-MHz sine-wave input, careful design is necessary to obtain a resolution of 12 bits or more.

For readers who want to find out more about jitter, please refer to a tutorial paper [33].

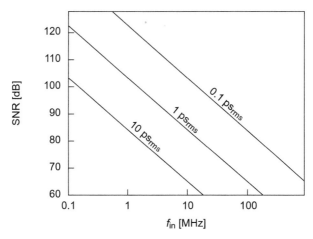

Figure 2.14   SNR degradation due to jitter.

---

[6]Photonic A/D converters have been proposed, where a femtosecond pulsed laser is used for sampling [32]. Jitter can be dramatically reduced by one order of magnitude or more compared with electric circuits.

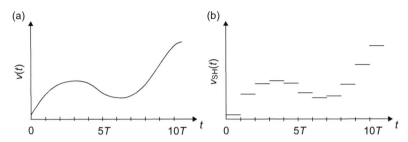

Figure 2.15 (a) Analog input signal and (b) sample-and-hold (S/H) signal.

2.1.4 Sample-and-hold (S/H) Signal

A signal that holds the sampled value until the next sampling time, as shown in Figure 2.15, is called a sample-and-hold (S/H) signal. For example, the output of an ideal D/A converter has such a waveform. This waveform is represented by Equation (2.5) in Subsection 2.1.1, when the pulse width τ is equal to the sampling period T. Its Laplace transform is written as

$$F_{\text{SH}}(s) = \frac{1 - e^{-Ts}}{s} \sum_{n=0}^{\infty} f(nT) e^{-nTs}. \tag{2.35}$$

This equation is considered to be Equation (2.7) representing a series of infinitely narrow pulses mentioned in Subsection 2.1.1, multiplied by the transfer function

$$H_{\text{SH}}(s) \equiv \frac{1 - e^{-Ts}}{s}. \tag{2.36}$$

The frequency characteristic of the S/H signal can then be expressed as

$$
\begin{aligned}
F_{\text{SH}}(j\omega) &= \frac{1 - e^{-j\omega T}}{j\omega} \sum_{n=0}^{\infty} f(nT) e^{-jn\omega T} \\
&= H_{\text{SH}}(s) F^*(j\omega).
\end{aligned} \tag{2.37}
$$

Here,

$$H_{\text{SH}}(j\omega) \equiv \frac{1 - e^{-j\omega T}}{j\omega} = T e^{-j\omega T/2} \frac{\sin\left(\frac{\omega T}{2}\right)}{\frac{\omega T}{2}}. \tag{2.38}$$

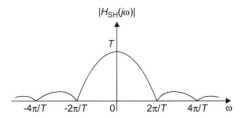

Figure 2.16   Sample-and-hold response.

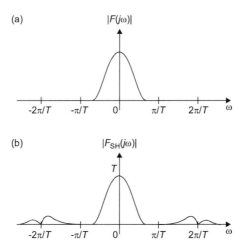

Figure 2.17   Spectra (a) before and (b) after sample-and-hold.

H_{SH} represents the sinc-filter characteristics as shown in Figure 2.16. The spectrum of the S/H signal is then

$$F_{\mathrm{SH}}(j\omega) = Te^{-j\omega T/2}\frac{\sin\left(\frac{\omega T}{2}\right)}{\frac{\omega T}{2}}\sum_{k=-\infty}^{\infty}F\left(j\omega - j\frac{2\pi k}{T}\right). \qquad (2.39)$$

This shows that the spectrum of the S/H signal is obtained by multiplying the periodically repeated spectrum of the original signal by the sinc-filter characteristic. The result is shown in Figure 2.17. It should be noted that as the frequency increases, the signal component in the main lobe attenuates because of the sinc filter. To compensate for the attenuation, it is necessary for the input signal to be multiplied by a transfer function having the inverse characteristic of the sinc function.

COLUMN: Sampling and the uncertainty relation

According to quantum mechanics, it is not possible to exactly determine the position and momentum of a particle at the same time. This is called the uncertainty relation between position and momentum. There is a similar uncertainty relation between energy and time: It takes an infinite time to measure the energy without any uncertainty. In other words, there exist errors in the energy measured within a short period. The uncertainty relation between energy and time, ΔE and Δt, is described as

$$\Delta E \Delta t \geq h. \tag{2.40}$$

Here, h is the Planck constant (6.6×10^{-34} Js).

If expressed in the words used in the A/D converter, the error in energy measurement increases as the sampling period becomes short. Since the energy is related to the voltage amplitude, the shorter the sampling period, the bigger the error in the sampled voltage. Thus, "noise" accompanies an ultrashort sampling period.

If half the sampling period is used for sampling, and the other half is used for A/D conversion, then $\Delta t = 1/(2f_s)$. The noise power ΔP due to the uncertainty is equal to $\Delta E \times f_s (\geq 2hf_s^2)$. When one wants to obtain a signal-to-noise ratio of 100 dB ($= 10^{10}$) at 100 GS/s, the required signal power is $2 \times 6.6 \times 10^{-34} \times (10^{11})^2 \times 10^{10} = 132$ mW. This should be compared with the required signal power that is calculated based on the thermal noise. For the same bandwidth and the SNR, the required signal power is $4.1 \times 10^{-21} \times 10^{11}/(2 \times 10^{10}) = 2$ W, where $kT = 4.1 \times 10^{-21}$ J and a bandwidth of $f_s/2$ are assumed. Thus, the effect of thermal noise is more significant than that of the uncertainty relation. However, while the thermal noise increases in proportion to the bandwidth, the noise caused by the uncertainty relation increases quadratically. Therefore, the influence of the latter could not be ignored if the sampling frequency increases up to a THz range in the future.

2.2 Quantization

Sampling discretizes continuous signals in the time domain, whereas quantization discretizes continuous physical quantities in the physical-quantity domain; for example, the voltage is quantized in the voltage domain. Consider the ramp wave V_{in} shown in Figure 2.18(a). After quantization, V_1 is obtained, which is represented by an N-bit digital code, $D_1 D_2 \cdots D_N$. In other words, V_1 shown here is the output of an N-bit D/A converter, whose input is

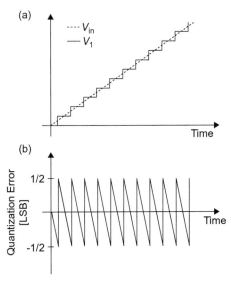

Figure 2.18   (a) Quantization of a ramp wave and (b) quantization error represented in LSB.

$D_1 D_2 \cdots D_N$. The quantization yields the quantization error V_Q that is the difference between the quantized value and the input analog value defined as

$$V_Q = V_1 - V_{in}. \tag{2.41}$$

Figure 2.18(b) shows the quantization error in LSB units. The quantization error is uniquely determined for a certain input value. Since the quantization error is an amount uniquely determined by the input, this model is called a deterministic model. The root mean square (rms) of the quantization error $V_{Q(rms)1}$ is expressed as

$$
\begin{aligned}
V_{Q(rms)1} &= \left[\frac{1}{T} \int_{-T/2}^{T/2} V_Q^2 dt \right]^{1/2} \\
&= \left[\frac{1}{T} \int_{-T/2}^{T/2} V_{LSB}^2 \left(\frac{-t}{T} \right)^2 dt \right]^{1/2} \\
&= \left[\frac{V_{LSB}^2}{T^3} \left(\frac{t^3}{3} \right)_{-T/2}^{T/2} \right]^{1/2} = \frac{V_{LSB}}{\sqrt{12}}.
\end{aligned} \tag{2.42}
$$

In general, it is not practical to analyze the quantization error that depends on the input signal because input signals are not predetermined.

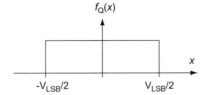

Figure 2.19 Probability density function for quantization errors.

Therefore, instead of the deterministic model, a stochastic model is introduced, where the quantization error does not depend on the input signal, but it is considered as a random variable defined by a distribution function. If the distribution function is constant as shown in Figure 2.19, the average quantization error $V_{Q(avg)}$ is

$$V_{Q(avg)} = \int_{-\infty}^{\infty} x f_Q(x) dx$$

$$= \frac{1}{V_{LSB}} \int_{-V_{LSB}/2}^{V_{LSB}/2} x dx = 0. \tag{2.43}$$

The root mean square of quantization error $V_{Q(rms)2}$ is

$$V_{Q(rms)2} = \left[\int_{-\infty}^{\infty} x^2 f_Q(x) dx \right]^{1/2}$$

$$= \left[\frac{1}{V_{LSB}} \int_{-V_{LSB}/2}^{V_{LSB}/2} x^2 dx \right]^{1/2} = \frac{V_{LSB}}{\sqrt{12}}. \tag{2.44}$$

Note that the root-mean-square value agrees with that obtained by Equation (2.42). Since in this model, the quantization error is dealt with stochastically, the quantization error is often referred to as the quantization "noise."

As mentioned above, in a strict sense, the quantization error is a value determined by the input, and it is not random noise. In other words, the stochastic model is an approximation. The number of discrete levels should be large enough for this approximation to be valid; typically, 6 bits or more is desirable. Also, the sampling number should be so large that all the possible digital codes appear with almost equal probability. Finally, there should be no correlations between the quantization errors of adjacent sampled values. The quantization noise model is shown in Figure 2.20, where $V_Q(nT)$ is the quantization noise.

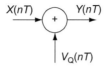

Figure 2.20   Model of quantization.

It is convenient to use the ratio of the input signal power to the quantization error as a performance specification of the A/D conversion. It is often called the signal-to-noise ratio (SNR) by considering the quantization error as noise. It is sometimes called a signal-to-quantization-noise ratio (SQNR) to clearly show that the noise in question is the quantization noise. If the input signal V_{in} is a sinusoidal wave and its amplitude is $V_{\text{ref}}/2$, its average magnitude $V_{\text{in(rms)}}$ is $V_{\text{ref}}/2\sqrt{2}$. By taking the ratio to the noise magnitude of Equation (2.42), the SNR can be calculated as

$$\text{SNR (in dB)} = 20 \log_{10} \frac{V_{\text{in(rms)}}}{V_{\text{Q(rms)}}}$$

$$= 20 \log_{10} \frac{V_{\text{ref}}/2\sqrt{2}}{V_{\text{LSB}}/\sqrt{12}} = 20 \log_{10} \sqrt{\frac{3}{2}} 2^N$$

$$= 6.02N + 1.76. \tag{2.45}$$

Similarly, if the input signal V_{in} is a ramp wave and its peak-to-peak voltage is V_{ref}, its rms magnitude $V_{\text{in(rms)}}$ is $V_{\text{ref}}/\sqrt{12}$, and the SNR can be expressed as

$$\text{SNR (in dB)} = 20 \log_{10} \frac{V_{\text{in(rms)}}}{V_{\text{Q(rms)}}}$$

$$= 20 \log_{10} \frac{V_{\text{ref}}/\sqrt{12}}{V_{\text{LSB}}/\sqrt{12}} = 20 \log_{10} 2^N$$

$$= 6.02N. \tag{2.46}$$

The reason why the SNR is larger by 1.76 dB for a sinusoidal input than for a ramp input is that the sinusoidal signal power is larger than the ramp signal power if the amplitudes are the same. Usually, Equation (2.45) is used, but note that the SNR depends on the input waveform. From Equation (2.45),

$$N = \frac{\text{SNR} - 1.76}{6.02} \tag{2.47}$$

is obtained. N calculated in this way from the SNR is called the effective number of bits (ENOB), and is often used to represent the resolution of data converters.

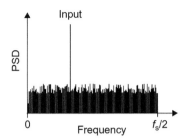

Figure 2.21 Power spectrum density (PSD) of quantization noise.

The spectrum of the A/D converter output for the sine-wave input is shown in Figure 2.21. The single peak at the input frequency represents the sinusoidal input before quantization, whereas quantization results in the noise floor almost uniformly distributed. The SNR can be estimated from such a spectrum, and hence the ENOB is evaluated by using Equation (2.47).

3

Basic Circuit Blocks

In this chapter, various circuits implementing sampling and quantization are presented. Called sample-and-hold (S/H) circuits and comparators, respectively, they are fundamental building blocks of data converters. First, a simple S/H circuit is described with effects of nonideal factors, such as on-resistance change, charge injection, and finite transition time, on the circuit performance. Next, practical circuit examples, as well as a bootstrap switch used for low-voltage operation, are presented. Then, thermal noise, power consumption, and sampling-time jitter are mentioned. As for the comparator, opamp-based comparators are first presented followed by multi-stage and latched comparators. Nonidealities due to imperfections in analog circuits and speed limitations are also described.

3.1 Sample-and-hold (S/H) Circuits

The sample-and-hold (S/H) circuit samples the continuously changing analog signal and temporarily holds the sampled value until the next sampling time. In a sense, it is a kind of analog memory circuits.

3.1.1 Basic Circuit

A basic S/H circuit is shown in Figure 3.1. This circuit consists of a switch S_1 and a hold capacitor C_{hold}, and operates in two modes. When S_1 is closed, the circuit is in the sample mode, whereas when S_1 is opened, it is in the hold mode. In the sample mode, the output voltage V_{out} follows the input voltage V_{in}. At the same time, the charge $Q(= C_{hold}V_{in})$ is stored in the capacitor C_{hold}. In the hold mode, the output is disconnected from the input and the charge equivalent to the input voltage V_{in1} immediately before the transition from the sample to hold mode is stored in the capacitor C_{hold}. Therefore, the output voltage during the hold mode is kept constant at $V_{out} = V_{in1}$.

41

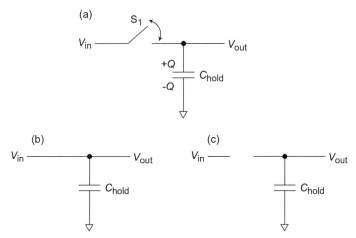

Figure 3.1   (a) Simple sample-and-hold (S/H) circuit and that in (b) the sample mode and (c) the hold mode.

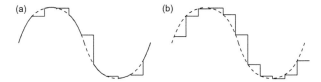

Figure 3.2   (a) Track-and-hold (T/H) and (b) sample-and-hold (S/H) waveforms.

This continues until the next sample mode. The output waveform is shown in Figure 3.2(a). In this circuit, the output tracks the input in the sample mode. Therefore, this mode is also called the track mode, and the circuit is accordingly called a track-and-hold (T/H) circuit. The operation mode without the tracking mode is called the sample-and-hold operation in a narrow sense, the output waveform of which is shown in Figure 3.2(b). However, unless there is a possible misunderstanding, the word sample-and-hold (S/H) is used for both circuits in the following explanation.

As shown in Figure 3.3, the actual circuit uses a MOSFET as a switch, which is turned ON and OFF by using the clock signal ϕ_S applied to the gate. When the MOSFET is an n-channel MOSFET, the circuit is in the sample mode if ϕ_S is HIGH, while it is in the hold mode if ϕ_S is LOW. A unity gain buffer is used as the output stage of the S/H circuit. This buffer plays a vital role of not only transmitting the input signal to the output but also supplying energy to drive the output load. For the output voltage to change in

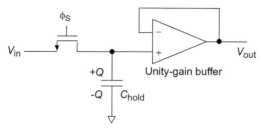

Figure 3.3 Sample-and-hold (S/H) circuit with a unity-gain buffer.

response to the input signal, a certain amount of energy is necessary, and if it is supplied directly from the input terminal, then the sampled values might be affected by the output load. Adding the buffer makes it possible to hold the input signal accurately without being affected by the output load. For the same reason, a unity gain buffer is also used on the input side, as is described in Subsection 3.1.4.

There is another sampling method called serial sampling [34]. The circuit diagram and the timing chart are shown in Figure 3.4. ϕ_1 and ϕ_2 are two-phase non-overlapping clock signals[1]. ϕ_{1a} means that it goes LOW slightly earlier than ϕ_1. When ϕ_1 is HIGH, the circuit is in the sample mode, and the output V_{out} is reset to V_{DD}. When ϕ_2 is HIGH, the circuit is in the hold mode, and the output is equal to $V_{DD} - V_{in}$. If there exists parasitic capacitance C_p associated with the output node, charges can move between C_{hold} and C_p in the transition from the sample mode to the hold mode, so an error occurs in the sampled value. Also, at the same time, the output changes from V_{DD} to $V_{DD} - V_{in}$. Therefore, the settling time must be taken into account.

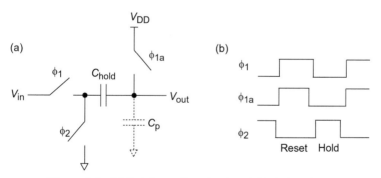

Figure 3.4 (a) Serial sampling circuit and (b) timing chart.

---

[1]They alternately repeat HIGH and LOW without becoming HIGH at the same time.

Table 3.1   Parallel sampling and serial sampling

	Parallel Sampling	Serial Sampling
Operation mode	track and hold	reset and hold
Input coupling	dc coupling	capacitive coupling
Charge injection amount	input dependent	input independent (with ϕ_{1a})
Input high-frequency feedthrough	attenuation due to C_{hold}	no attenuation
Influence of C_p	small (charged in track mode)	large
Settling time	short (due to tracking)	long (due to reset)

The sampling method described earlier is sometimes called parallel sampling to distinguish it from serial sampling. A comparison of both is given in Table 3.1. Serial sampling has the advantage that the charge injection does not depend on the input signal magnitude. These details are described below (Subsection 3.1.3.2).

3.1.2 Output Waveform of S/H Circuit

Consider the output waveform of a S/H circuit shown in Figure 3.2(a) in detail. A typical waveform obtained from a practical circuit is shown in Figure 3.5. When ϕ_S shown in Figure 3.3 changes from HIGH to LOW, the S/H circuit transfers from the sample mode to the hold mode. The MOSFET switch eventually turns off in this transition, but not instantaneously. The MOSFET continues to be conductive for a short time until all the charges accumulated in the channel disappear. This time delay is called an aperture time. Then, after another certain amount of time, called a settling time, the circuit settles into the hold mode. There exists a slight difference between

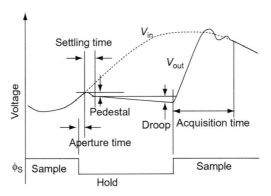

Figure 3.5   Output waveform of sample-and-hold (S/H) circuit.

the input and output values, which is called pedestal. Once the output voltage has been held, the output voltage can still change slightly over time. This is because of charge leakage paths existing in the actual device. This phenomenon is called droop. Although not shown in this figure, even when the MOSFET is in the OFF state, there is a parasitic capacitance connecting the source with the drain. The input signal may leak into the output through this capacitive coupling. This is called input feedthrough. Likewise, the clock signal applied to the gate may leak to the output, which is called clock feedthrough.

When ϕ_S goes from LOW to HIGH, and the circuit shifts from the hold mode to the sample mode, the output completely follows the input after a certain time called an acquisition time. In this period, the hold capacitance C_{hold} is charged or discharged through the on-resistance of the MOSFET. This is a transient response phenomenon that can be described with the well-known CR time constant. As shown in this figure, overshoot may occur during this transition.

Consider a simplified model shown in Figure 3.6, and suppose that the hold mode is terminated at $t = 0$. Since the output asymptotically approaches to the final value exponentially, there exists a settling error. Table 3.2 indicates the elapsed time (t_1) required to reduce the settling error from 1% to 0.01%. Here, R_{on} is the resistance when the MOSFET used for the switch is in the ON state (on-resistance). If an error of less than 0.01% is required, it is necessary to wait for more than nine times the CR time constant. The settling error needs to be less than the quantization error of an A/D converter.

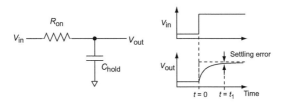

Figure 3.6 Settling behavior in sample-and-hold (S/H) circuit.

Table 3.2 Settling error

Settling error at t_1	t_1 [$R_{\text{on}}C_{\text{hold}}$]
1%	4.6
0.1%	6.9
0.01%	9.2

For example, when 10-bit resolution is required, it is necessary to set the settling error to be less than 0.1%, which means that it takes more than seven times the time constant to settle.

A small CR time constant is desirable for high-speed operation. To this end, it is then necessary to reduce either the on-resistance or the holding capacitance. It seems useful to widen the gate to reduce the on-resistance. However, the junction capacitance of the MOSFET increases accordingly, which can deteriorate circuit operation speed. If the hold capacitance is reduced, the effects of thermal noise and process-induced variations become large. Therefore, there is a lower limit on the practical capacitance value. In actual design, detailed circuit-level simulations are necessary.

3.1.3 Nonideal Factors

The function of the S/H circuit is simple, but there are several critical nonideal factors to consider in implementing a practical circuit. In this subsection, the impact of nonidealities on the S/H circuit performance are described. Nonideal factors considered here are the on-resistance change of the switch MOSFET, the charge injection when the MOSFET switch turns off, and the finite transition time from the sample mode to the hole mode.

3.1.3.1 Change in on-resistance

Consider the on-resistance of the MOSFET switch in the S/H circuit shown in Figure 3.3. Since the drain voltage and the source voltage of the MOSFET are almost the same, the MOSFET operates in the linear region. Therefore, based on a simple MOSFET model, the drain current I_D is expressed as

$$I_D = \mu C_{ox} \frac{W}{L} \left[(V_{GS} - V_{tn}) V_{DS} - \frac{1}{2} V_{DS}^2 \right]. \tag{3.1}$$

Here, μ and C_{ox} is the electron mobility and the channel capacitance per unit area. V_{tn} is the threshold voltage of the n-channel MOSFET. The on-resistance R_{on} is then obtained as

$$R_{ON} = \left(\frac{\partial I_D}{\partial V_{DS}} \right)^{-1} = \left(\mu C_{ox} \frac{W}{L} (V_{GS} - V_{tn} - V_{DS}) \right)^{-1}$$
$$\approx \left(\mu C_{ox} \frac{W}{L} (V_{GS} - V_{tn}) \right)^{-1} \tag{3.2}$$

When the input voltage V_{in} increases, the gate-source voltage V_{GS} $(= V_{DD} - V_{in})$ decreases, and the on-resistance increases. This means that

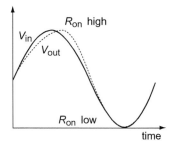

Figure 3.7 Signal distortion due to on-resistance change.

the CR time constant also increases as V_{in} increases, which results in a long delay time. Therefore, the output signal is distorted as shown in Figure 3.7. Increasing W/L and the clock voltage V_{DD} or decreasing the threshold V_{tn} is useful for reducing such a change in the on-resistance.

In the circuit shown above, $V_{GS} > V_{tn}$, or $V_{DD} - V_{tn} > V_{in}$, is assumed. However, the input voltage amplitude should be as large as possible to widen the dynamic range. To extend the upper limit on the input voltage from $V_{DD} - V_{tn}$ to V_{DD}, a transmission gate shown in Figure 3.8(a) is widely used as a switch[2]. Figure 3.8(b) shows the on-resistance change of the transmission gate. Here, V_{tp} is the threshold voltage of the p-channel MOSFET. The on-resistance of the transmission gate is a combined resistance of the nMOSFET and the pMOSFET. Therefore, the change in the on-resistance is small compared with the case where only one of them is used. Nevertheless, as shown in this figure, the on-resistance tends to increase around the mid-region between 0 and V_{DD}, where the delay time is slightly longer compared

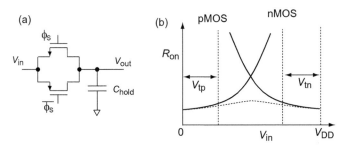

Figure 3.8 (a) S/H circuit using a transmission gate and (b) on-resistance. The dotted line is a combined resistance.

---

[2]This is also called a pass gate.

with that in the regions at both ends, and the signal might be distorted. Even if it is not a severe problem in the digital circuit, this can result in a significant distortion in analog circuits as shown below.

Suppose that the input is expressed as

$$V_{\text{in}}(t) = V_0 + V_0 \cos [\omega_{\text{in}} t], \tag{3.3}$$

then the output is obtained as

$$\begin{aligned} V_{\text{out}}(t) &\approx V_0 + V_0 \cos \left[\omega_{\text{in}} t - \tan^{-1}(R_{\text{on}} C_{\text{hold}} \omega_{\text{in}})\right] \\ &\approx V_0 + V_0 \cos [\omega_{\text{in}} t] + V_0 R_{\text{on}} C_{\text{hold}} \omega_{\text{in}} \sin \omega_{\text{in}} t. \end{aligned} \tag{3.4}$$

The approximation is made by assuming $R_{\text{on}} C_{\text{hold}} \omega_{\text{in}} \ll 1$. If the input is a full-scale sine wave that varies between 0 and V_{DD}, the input signal makes a round trip between 0 and V_{DD} of Figure 3.8(b) during one period. Therefore, R_{on} changes with an angular frequency of $2\omega_{\text{in}}$. If it is written as

$$R_{on}(t) = R_0 + R_1 \cos [2\omega_{\text{in}} t] + R_2 \cos [4\omega_{\text{in}} t] + \dots, \tag{3.5}$$

then

$$\begin{aligned} V_{\text{out}}(t) &\approx V_0 + V_0 \cos [\omega_{\text{in}} t] \\ &\quad + V_0 \left(R_0 + R_1 \cos [2\omega_{\text{in}} t] + R_2 \cos [4\omega_{\text{in}} t] + \dots \right) \\ &\quad \times C_{\text{hold}} \omega \sin \omega_{\text{in}} t \end{aligned} \tag{3.6}$$

is obtained. The amplitude of the third harmonic component is then $V_0 \omega_{\text{in}} C_{\text{hold}} R_1/2$. For example, if it is attempted to suppress this third-order distortion to be less than 60 dB,

$$\frac{\omega_{\text{in}} C_{\text{hold}} R_1}{2} < 10^3 \tag{3.7}$$

is required. From this equation, to keep the distortion below a specified level, the input signal bandwidth is limited by the variation in the on-resistance, or by the value of R_1 [35].

3.1.3.2 Charge injection

The MOSFET switch in Figure 3.3 turns off in the transition from the sample mode to the hold mode. At that time, carriers accumulated in the channel flow out of the MOSFET through the source and drain terminals. Some of these carriers flow into the hold capacitor C_{hold}, and accordingly the sampled

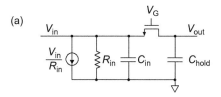

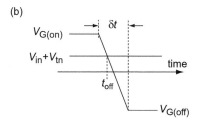

Figure 3.9 (a) S/H circuit model for charge injection analysis and (b) V_G waveform.

voltage changes. In a circuit using an n-channel MOSFET switch, carriers are electrons, which cancels partially the positive charges accumulated in the capacitor. Therefore, V_{out} decreases, resulting in the pedestal shown in Figure 3.5.

A S/H circuit model to analyze the effect of charge injection is shown in Figure 3.9. Let us assume that the gate voltage changes by ΔV_G from $V_{G(on)}$ to $V_{G(off)}$ in a finite time of δt. The MOSFET turns off when $V_G - V_{in} = V_{tn}$ or at t_{off} in Figure 3.9(b). The charge Q_{ch} accumulated in the MOSFET channel, when it turns on, can be written as

$$Q_{ch} = WLC_{ox}(V_{G(on)} - V_{tn}). \tag{3.8}$$

Figure 3.10 shows an example of evaluating the ratio of the charge ΔQ_S injected into the hold capacitor to the total channel charge Q_{ch} just before the MOSFET turns off [10]. Here, B is the switching parameter defined as

$$B = V_{ov}\sqrt{\frac{\mu C_{ox}W/L}{|\alpha|C_{hold}}}. \tag{3.9}$$

V_{ov} is the MOSFET overdrive voltage $V_{GS} - V_{tn}$, and $\alpha = \Delta V_G/\delta t$.

Figure 3.10 shows that if B is small, $\Delta Q_S/Q_{ch} = 0.5$, which means that the charges flowing out of the source and drain are the same. Half of the charge accumulated in the channel is injected into the hold capacitor. The

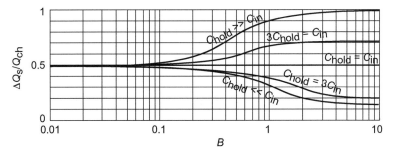

Figure 3.10 Ratio of the charge ΔQ_S injected into the holding capacitor to the channel charge Q_{ch} as a function of the switching parameter B defined by Equation (3.9) [10].

sampled voltage then changes by ΔV, which can be written as

$$\Delta V = \frac{\Delta Q_{hold}}{C_{hold}} = \frac{Q_{ch}}{2C_{hold}} = -\frac{C_{ox}LWV_{ov}}{2C_{hold}}. \tag{3.10}$$

Remember that the overdrive voltage, $V_{ov} = V_{GS} - V_{tn} = V_{DD} - V_{in} - V_{tn}$, depends on the input voltage V_{in}. Therefore, ΔV is not a constant but depends on the input voltage V_{in}. This means that distortion occurs in the output signal.

As the switching parameter B increases, the amount of charge flowing into the source is not equal to that to the drain. If $C_{in} \gg C_{hold}$, then the charge injection into the hold capacitor can be suppressed. However, since a large input capacitance limits the input signal bandwidth, this is not a practical solution when high-speed operation is required. Further, although it may be considered to reduce the channel width W to suppress the charge injection, the on-resistance increases and the signal band becomes narrow. Conversely, if a MOSFET with a wide channel is used as a switch, the signal band can be widened, but the charge injection is enhanced, and the distortion is also increased.

Adding a dummy gate as shown in Figure 3.11 is proposed to suppress charge injection effects. It is intended to absorb charges flowing out of M_1 by turning on the dummy gate M_2. However, it is difficult to turn off and on these MOSFETs simultaneously, because the on and off timing depend on the input signal magnitude if the clock transient time is not zero. Also, with the transmission gate shown in Figure 3.8, it is possible to some extent to suppress the influence of charge injection by adjusting the gate widths of p/n-channel MOSFETs. However, the channel charge and the turning-off timing

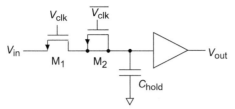

Figure 3.11 Dummy gate (M_2) to suppress charge injection effects.

of these MOSFETs depend on V_{in}. Therefore, it is difficult to suppress the charge injection altogether.

3.1.3.3 Nonzero transition time

When the transition time of the clock voltage driving the MOSFET switch is not zero, the sampling time in the S/H circuit shown in Figure 3.3 varies depending on the magnitude of the input signal. This is shown in Figure 3.12. The negative edge, where V_{clk} changes from HIGH to LOW, determines the sampling time. More specifically, the sampling occurs at the time when $V_{in} + V_{tn} = V_{clk}$ on the negative edge. Sampling times are indicated by t_1, t_2, and t_3, which are different from ideal sampling times of t_1^0, t_2^0, and t_3^0. Deviations in the sampling times are as shown by $\delta t_1, \delta t_2$, and δt_3. If the input signal is large, the sampling timing is advanced, and if the input signal is small, it is delayed. In other words, this jitter depends on the input signal magnitude. Therefore, the signal-to-noise ratio deteriorates as was described in Subsection 2.1.3. A clock signal with sharp negative edges is useful to suppress this effect, and it can be effective to use scaled-down high-speed MOSFETs or bipolar junction transistors (BJTs) having a high transconductance g_m.

3.1.4 Circuit Examples

In the following, closed-loop circuits using operational amplifiers (opamps) and open-loop circuits suitable for high-speed operation will be described

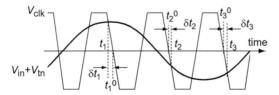

Figure 3.12 Effect of finite clock transition time on sample timing.

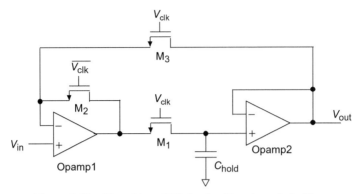

Figure 3.13   Closed-loop S/H circuit with unity-gain buffers.

as examples of the S/H circuits in which the nonideal factors are taken into consideration.

3.1.4.1 Closed-loop S/H circuits

An example of closed-loop S/H circuits [36] is shown in Figure 3.13. When V_{clk} is HIGH, the circuit is in the sample mode. Since the feedback is applied to the first-stage opamp (Opamp1) through M_3, it operates as a unity gain amplifier, so that the output voltage V_{out} is equal to V_{in}. Since the input impedance of Opamp1 is quite high, the input voltage is accurately captured by C_{hold}. When V_{clk} is LOW, the circuit is in the hold mode. Even if there is an offset in the output buffer opamp (Opamp2), the offset effect on the output is reduced by an inverse of the gain of Opamp1. Therefore, the second stage amplifier can be replaced with a simple source follower.

In the hold mode, M_2 turns on, and feedback is applied to Opamp1 to prevent its output from being fixed to the power supply voltage or ground. This allows the output to track the input voltage in a relatively short time at the beginning of the next sample mode. However, since charge injection in the transition from the sample mode to the hold mode depends on the input signal magnitude, it is difficult to reduce the distortion as is the case in Figure 3.3. Also, if the transition time of V_{clk} cannot be ignored, the sampling timing fluctuates depending on the input signal magnitude.

Another example of closed-loop S/H circuits aiming at solving these problems is shown in Figure 3.14 [36]. When V_{clk} is HIGH, the circuit is in the sample mode. If the gain of Opamp2 is sufficiently large, both the source and drain potentials of M_1 are zero regardless of the input voltage V_{in}. Therefore, even if charges are injection from M_1 in the transition from

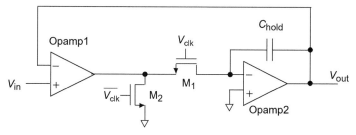

Figure 3.14 Low-distortion closed-loop S/H circuit.

the sample mode to the hold mode, the amount of injected charge is constant. Therefore, the offset caused by M_1 is independent of the input signal magnitude, resulting in no distortion. Also, the sampling time is not signal dependent, and the sampling occurs at virtually ideal intervals. M_2 forces the output of Opamp1 to be fixed to zero in the hold mode, allowing a fast transition from the hold mode to the sample mode. At the same time, the input feed-through can be suppressed because the leakage signal from the input escapes to the ground through M_2 in the hold mode. On the other hand, note that the requirement for stable operation still limits the operating speed.

3.1.4.2 Open-loop S/H circuits

Open-loop S/H circuits are used to achieve high-speed operation. A typical example is shown in Figure 3.15 [37]. The circuit uses a diode bridge consisting of D_1, D_2, D_3, and D_4 with the same size. When V_{clk} is HIGH, the circuit is in the sample mode. All the diodes turn on, and the forward current flows through them. Potential differences across the diodes are the

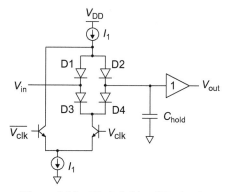

Figure 3.15 Diode-bridge S/H circuit.

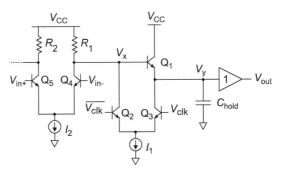

Figure 3.16   Switched emitter-follower S/H circuit.

same, and V_{out} follows V_{in}. When V_{clk} is LOW, the circuit is in the hold mode. No current flows through the diodes, and all diodes turn off. Then the output is disconnected from the input. The circuit can be modified for higher-speed operation [38]. If pn-junction diodes are used as D_1 to D_4, the lifetime of minority carriers limits the maximum operating speed. Using Schottky diodes, which operate only with majority carriers, can improve the speed performance.

In the diode-bridge S/H circuit mentioned above, a number of pn junctions exist between the power supply and ground, if those in BJTs are also taken into account. Thus, a relatively high supply voltage is required to turn on all of them simultaneously. S/H circuits consisting of only BJTs were proposed to reduce the supply voltage. Figure 3.16 shows a schematic diagram of a circuit called a switched emitter follower [39]. A differential circuit is often used, but a half circuit is shown here for simplicity. When V_{clk} is HIGH, Q_1 and Q_3 turn on and operate as an emitter follower. V_{out} tracks V_x with the voltage drop across the pn diode between the emitter and base of Q_1. When V_{clk} goes LOW, the current I_1 is steered towards Q_2 and flows through R_1. Then, V_x drops by $I_1 \cdot R_1$ and $V_x - V_y$ becomes less than the threshold voltage of Q_1. Therefore, Q_1 turns off, V_x and V_y are separated from each other, and the circuit is in the hold mode. This circuit is called a switched emitter follower because V_{clk} turns on and off the emitter follower.

Let us consider whether or not the sampling time changes with the input voltage. As mentioned above, the voltage drop due to R_1 turns off Q_1. If the output impedance of Q_4 is infinitely large, the Q_4 collector current does not change when the emitter to collector voltage of Q_4 changes as V_x decreases. Therefore, even if V_{in-} changes, the voltage across R_1 is not affected, and the sampling timing does not change either. What happens if the output impedance of Q_4 is finite? When V_x drops in the transition to the hold

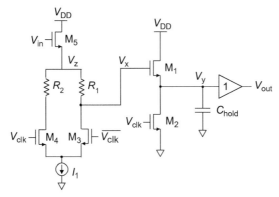

Figure 3.17 Switched source-follower S/H circuit.

mode, the emitter-collector voltage of Q_4 decreases, and at the same time, the Q_4 collector current also decreases. This current reduction counteracts the decrease in V_x. The amount of current reduction is greater for higher collector currents than for lower collector currents. Thus, for a large $|V_{in-}|$, the sampling time is more delayed compared with a small $|V_{in-}|$.

This can be a problem when the circuit is implemented by using scaled-down MOSFETs instead of BJTs because it is difficult to obtain a high output impedance in miniaturized MOSFETs. The sampling timing thus depends on the input magnitude, and as a result, the sampled signal is distorted. Figure 3.17 shows a proposed circuit to solve this problem [40, 41]. This S/H circuit is called a switched source follower. In this circuit, even if the output impedance of M_3 is finite, the same constant current I_1 always flows in R_1. Therefore, when the circuit turns into the hold mode, the voltage drop across R_1 does not change, and the sampling time is not affected by the input signal. However, if the output impedance of M_5 is finite, the voltage between the gate and source of M_5 may change depending on the value of V_{in}, so careful design is required.

3.1.5 Bootstrap Switch

In this subsection, a switch used in low-voltage S/H circuits, called a bootstrap switch, is described[3].

---

[3]This is named after the story that when Baron Munchausen fell into a swamp, he escaped from it by pulling his bootstrap (it is said that he pulled his hair instead of the bootstrap in the original story). Here, it means a method to generate a higher voltage than the supply voltage by itself.

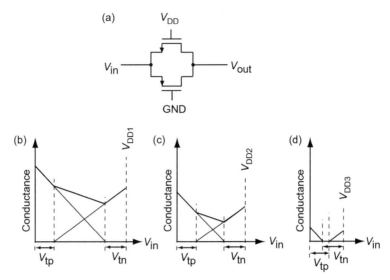

Figure 3.18   (a) Schematic of a pass gate, and ON conductance under (b) standard, (c) low, and (d) very low supply voltages.

Figure 3.18 shows the on-conductance of the pass gate for three supply voltages that decrease in the order of V_{DD1}, V_{DD2}, and V_{DD3}. The on-conductance of the pass gate G_{on} is represented as

$$G_{on} = \mu_n C_{ox} \left[\frac{W}{L}\right]_n V_{ov,n} + \mu_p C_{ox} \left[\frac{W}{L}\right]_p V_{ov,p}. \tag{3.11}$$

Here, the overdrive voltages of each MOSFET are

$$V_{ov,n} = V_{DD} - V_{in} - V_{tn} \tag{3.12}$$

$$V_{ov,p} = V_{in} - V_{tp}. \tag{3.13}$$

V_{tn} and V_{tp} are the threshold voltages of the n-channel and p-channel MOSFETs, respectively. If

$$\mu_n \left[\frac{W}{L}\right]_n = \mu_p \left[\frac{W}{L}\right]_p \tag{3.14}$$

is satisfied, then

$$G_{on} = \mu_n C_{ox} \left[\frac{W}{L}\right]_n (V_{DD} - V_{tn} - V_{tp}) \tag{3.15}$$

is obtained. Therefore, as shown in Figure 3.18(b) and (c), the conductance decreases with the decrease in V_{DD}. If $V_{DD} < V_{tn} + V_{tp}$, the conductance is 0 near the center of the signal amplitude (Figure 3.18(d)), which means that the switch never turns on.

If the voltage higher than the sum of the input voltage and the threshold voltage can be generated in the circuit and supplied to the gate, the switch MOSFET remains to turn on even with a low supply voltage. The concept of implementing this is shown in Figure 3.19. In Figure 3.19(a), the voltage higher than the input signal V_{in} by V_{DD} is generated and applied to the gate of M_{11} as V_{clk}. If $V_{DD} \geq V_{tn}$, M_{11} turns on and the input voltage can be sampled. Therefore, this circuit can operate at considerably lower voltages than circuits using a conventional pass gate, where $V_{DD} \geq V_{tn} + V_{tp}$ is required.

One of the important consequences that can be seen from Figure 3.18 is as follows: Use an n-channel MOSFET when connecting to ground, and use a p-channel MOSFET when connecting to V_{DD}. When connecting to V_{DD} or ground, a pass gate is not necessary. In contrast, the pass gate should be used when connecting to the signal that can vary from ground to V_{DD}.

Because it is not practical to include a voltage source in an actual integrated circuit, V_{DD} is generated by a charged capacitor C_b as shown in Figure 3.19(b). As shown in Figure 3.19(c), M_3 and M_{12} are required to charge C_b by connecting the capacitor to the power supply and ground. In the sample mode, by turning off M_3 and M_{12} and turning on M_8 and M_9, a voltage higher than the signal voltage by V_{DD} is applied to the M_{11} gate. In the hold mode, on the other hand, M_{10} is turned on and M_8 is turned off so that M_{11} is turned off.

Figure 3.20(a) shows a circuit in which M_8, M_{10}, and M_{12} in Figure 3.19(c) are replaced by MOSFETs. As shown in Figure 3.18(d), it is

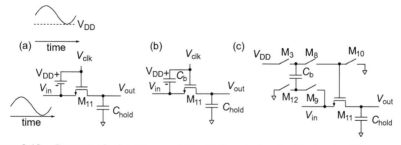

Figure 3.19   Concept of a bootstrap switch. (a) Target circuit, (b) circuit with a capacitor, and (c) circuit with switches for charging the capacitor.

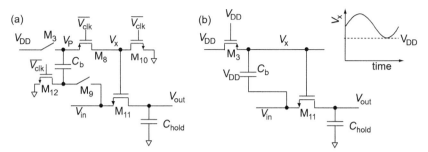

Figure 3.20   (a) Bootstrap switch circuit using MOSFETs and (b) circuit in the sample mode.

necessary to use an nMOSFET as a switch to connect to ground and a pMOSFET to connect to V_{DD}. In this case, then, a pMOSFET is used for M_8, and nMOSFETs are used for M_{10} and M_{12}. When V_{clk} is LOW, M_{10} turns on and V_x is connected to ground so that M_{11} turns off and the circuit is in the hold mode. Since M_{12} also turns on at this time, C_b can be charged if M_3 is set to on.

When V_{clk} becomes HIGH, V_x switches from ground to V_P. If M_3 and M_{12} turn off and M_9 turns on, a voltage higher than V_{in} by V_{DD} is supplied to the gate of M_{11}, and the circuit is in the sample mode. Let us consider how high the gate voltage of M_3 should be at this time. Since M_3 is connected to V_{DD}, it must be a pMOSFET. In the sample mode, this pMOSFET should be turned off. Since V_P is higher than V_{DD} in the sample mode as shown in Figure 3.20(b), M_3 cannot be turned off if its gate voltage is V_{DD}. On the other hand, to periodically charge C_b, it is necessary to turn on M_3 in the hold mode. Therefore, the gate of M_3 needs to be connected to the boosted node V_x like M_{11}. Also, since M_9 needs to be turned on in the range from 0 V to V_{DD}, an nMOSFET is used, and it is necessary to apply a voltage higher than V_{DD} to the gate. So, the gate of this MOSFET also needs to be connected to the boosted node V_x. Thus a circuit diagram considering these is shown in Figure 3.21(a).

Since V_x is higher than V_{DD}, the reliability with respect to M_{10} in Figure 3.21(a) is an issue as the gate-to-drain voltage exceeds the allowable voltage determined by the process, or in the worst case, there is a possibility of damage. Inserting M_{14} as shown in Figure 3.21(b) [42] can prevent an excessive potential difference from occurring between the gate and drain of M_{10}, even if V_x becomes $2V_{DD}$ in the sample mode.

For readers interested in the content of this section, please refer to the articles describing the bootstrap switch [35, 43].

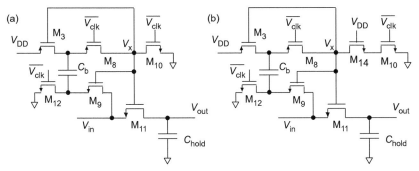

Figure 3.21 Practical bootstrap switch circuit.

3.1.6 Thermal Noise

In this subsection, the noise which limits the performance of S/H circuits is considered. In general, there are two kinds of noise: external noise having sources outside, and intrinsic noise caused by elements contained in the circuit. The former includes, for example, the noise coming from other circuits fabricated on the same substrate and the noise caused by fluctuations in the supply voltage and the ground potential. The latter includes thermal noise and 1/f noise (flicker noise) caused by statistical properties of electrons and traps. Here, the thermal noise is focused on because it is essential and inevitable[4]. Thermal noise determines the lower limit on the detectable signal magnitude. In other words, it determines the upper limit on the resolution in A/D conversion. Moreover, as described below, the magnitude of thermal noise, the operation speed, and the power consumption are closely related to one another. Therefore, in particular, in recent years there has been a growing interest in low power consumption circuit designs taking thermal noise into account.

The on-resistance of the MOSFET causes thermal noise in the S/H circuit. The power spectral density of the thermal noise generated by a resistor can be expressed as

$$\mathrm{PSD}(f) = 4kT, \tag{3.16}$$

---

[4]1/ noise is omitted here because it can be reduced by chopper stabilization or correlated double sampling method.

which is almost constant up to a THz range. The thermal noise power of interest can be evaluated by integrating this over a frequency range of $[f_1, f_2]$

$$P_\mathrm{n} = \int_{f_1}^{f_2} 4kT df = 4kT(f_2 - f_1) = 4kT\Delta f, \tag{3.17}$$

where $f_2 - f_1 = \Delta f$.

The thermal noise of a resistor R is represented by either an equivalent voltage source or an equivalent current source as shown in Figure 3.22. The equivalent voltage source is expressed as

$$\overline{v_\mathrm{n}^2} = P_\mathrm{n} R = 4kTR\Delta f, \tag{3.18}$$

while the equivalent current source is written as

$$\overline{i_\mathrm{n}^2} = \frac{P_\mathrm{n}}{R} = 4kT\frac{1}{R}\Delta f. \tag{3.19}$$

The S/H circuit in the sample mode is rewritten as shown in Figure 3.23(a). The on-state MOSFET is regarded as a noiseless resistor with the noise source indicated by Figure 3.22. In the hold mode, since the noise source is separated from the output, there is no influence of thermal noise on the output. Figure 3.23(b) shows the output waveform of the S/H circuit. The sampled values at $n - 1$, n, and $n + 1$ are different from one another due to the thermal noise. When looking at the noise generated in the circuit from the output terminal, the noise is called output-referred noise, and in this case, it is obtained as follows

$$\overline{v_\mathrm{out}^2} = \int_0^\infty 4kTR\left|\frac{1}{1 + j2\pi fRC}\right|^2 df$$

$$= 4kTR \cdot \frac{1}{4RC} = \frac{kT}{C}. \tag{3.20}$$

It seems strange that the output-referred noise does not depend on R. The explanation of this is as follows. $\overline{v_\mathrm{n}^2}$ in Equation (3.18) is proportional to R.

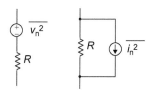

Figure 3.22 Thermal noise models for a resistor.

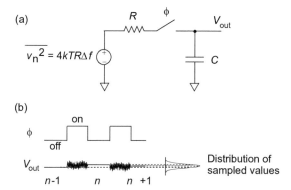

Figure 3.23 Thermal noise in a S/H circuit. (a) Circuit in which the switch MOSFET is replaced with a resistor with thermal noise and (b) output waveform.

The signal bandwidth is inversely proportional to R. Therefore, these cancel each other, and the output-referred noise does not depend on R. C exists in the expression, but note that the capacitor is not a noise source, but the noise is due to the resistor.

For example, based on Equation (3.20), the thermal noise becomes approximately 64 μV_{rms} for $C = 1$ pF. If the input full scale is 1 V, this corresponds to approximately 13-bit resolution, which gives a theoretical limit. The relation between the hold capacitor C_{hold} and the SNR is shown in Table 3.3.

Equation (3.20) can be derived from the equipartition law known in statistical physics. This means that the average energy per degree of freedom of a physical system in thermal equilibrium is given by $kT/2$. By applying this to the electrostatic energy stored in the capacitor,

$$\overline{\frac{1}{2}Cv_{out}^2} = \frac{1}{2}kT \qquad (3.21)$$

Table 3.3 Upper limit on SNR

SNR [dB]	C_{hold} [pF]
20	0.00000083
40	0.000083
60	0.0083
80	0.83
100	83
120	8300
140	830000

is obtained. This can be rewritten as

$$\overline{v_{\text{out}}^2} = \frac{kT}{C}.$$ (3.22)

In the S/H circuit shown in Figure 3.23, if the circuit time constant CR is small and if the noise is not correlated between adjacent samples, the thermal noise is white noise, which spreads between a frequency band of $[0, f_\text{s}/2]$. From Equation (3.20), the voltage noise per unit frequency after sampling can be expressed as

$$\frac{\overline{v_{\text{out}}^2}}{f_\text{s}/2} = \frac{2}{f_\text{s}} \frac{kT}{C}.$$ (3.23)

On the other hand, from Equation (3.18), the voltage noise per unit frequency generated by the resistor is

$$\frac{\overline{v_\text{n}^2}}{\Delta f} = 4kTR.$$ (3.24)

Taking the ratio of these yields

$$\frac{\overline{v_{\text{out}}^2}/f_\text{s}/2}{\overline{v_\text{n}^2}/\Delta f} = \frac{1}{2f_\text{s}} \frac{1}{CR} = \frac{T_\text{s}/2}{\tau}.$$ (3.25)

This means that the voltage noise per unit frequency increases by a factor of $\frac{T_\text{s}/2}{\tau}$ after the sampling. The reason for the increase after sampling is that the thermal noise initially spreading over a wide frequency range is folded into the narrow range by sampling.

3.1.7 Power Consumption

Let us consider the power consumption of the S/H circuit taking the thermal noise described above into account. The externally supplied power should be large enough that the output voltage corresponding to 1 LSB is larger than the output-referred noise. The circuit model for this analysis is shown in Figure 3.24.

As in the previous section, if only thermal noise of the on-resistance of the MOSFET switch is taken into consideration, the noise power v_n^2 can be written as

$$\overline{v_\text{n}^2} = \frac{kT}{C_{\text{hold}}}.$$ (3.26)

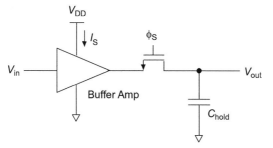

Figure 3.24 S/H circuit model for power consumption estimation.

Full-scale input signal power v_s^2 can be written as

$$v_\mathrm{s}^2 = \left(\frac{V_\mathrm{FS}}{2\sqrt{2}}\right)^2. \tag{3.27}$$

Here, V_FS is the full-scale voltage. The ratio between these is equivalent to the dynamic range D and can be written as

$$D = \frac{v_\mathrm{s}^2}{v_\mathrm{n}^2} = \frac{V_\mathrm{FS}^2 C_\mathrm{hold}}{8kT}. \tag{3.28}$$

From this equation, the holding capacitor C_hold required to obtain the dynamic range D can be obtained as

$$C_\mathrm{hold} = \frac{8kTD}{V_\mathrm{FS}^2}. \tag{3.29}$$

Furthermore, if the capacitor C_hold is charged in half the sampling period T_s using a constant current I_S, the required current can be estimated as

$$I_\mathrm{S} = \frac{C_\mathrm{hold} V_\mathrm{FS}}{T_\mathrm{s}/2}. \tag{3.30}$$

The corresponding power consumption $P_\mathrm{S/H}$ is then found as

$$P_\mathrm{S/H} = I_\mathrm{S} V_\mathrm{FS} = 16kT f_\mathrm{s} D. \tag{3.31}$$

Here, f_s is the sampling frequency, which is the inverse of T_s. Equation (3.31) indicates that the larger the required dynamic range, or the higher the resolution, the more power is required. It is also understood that high power consumption is required when sampling at high speed.

The above expression can also be written as

$$P_{S/H} = \frac{C_{hold}V_{FS}}{T_s/2}V_{FS} = 2C_{hold}f_sV_{FS}^2. \qquad (3.32)$$

Here, $V_{DD} = V_{FS}$ is assumed. Equation (3.32) is equivalent to the dynamic power consumption for CMOS digital circuits except for a coefficient of 2. The difference in the coefficient results from the fact that the circuit operates at half the sampling frequency.

For readers who are interested in further details, the references [44, 45] are listed at the end of the book.

3.1.8 Jitter

The effect of clock jitter on the S/H circuit was described in Subsection 2.1.3. Jitter means that the sampling time randomly fluctuates. Since inverters are commonly used as a clock buffer for supplying clock signals to MOSFET switches, let us consider the cause of jitter in the inverter.

When the supply voltage fluctuates, not only the output amplitude but also the slope changes. Therefore, as shown in Figure 3.25, fluctuation of the supply voltage ΔV_{DD} causes jitter Δt_{samp}. For the sake of simplicity, suppose that the cause of jitter is only the fluctuation of the supply voltage, and estimate its effect. For example, if a supply voltage varies by 10%, this causes a variation of the rising edge of approximately 5%. A rise time of 30–60 ps then corresponds to jitter of 1.5–3 ps. Speeding up the MOSFET with miniaturization can suppress jitter. In the digital CMOS circuit, when the input/output changes between "0" and "1", large current flows to the power supply line and the ground line, so the potentials of V_{DD} and ground fluctuate considerably. Therefore, sharing the ground and V_{DD} of the digital circuit on the same substrate should be strictly avoided.

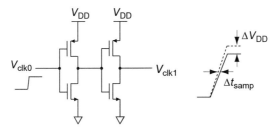

Figure 3.25   Jitter due to supply voltage fluctuation.

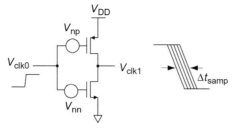

Figure 3.26   Jitter due to thermal noise.

Jitter is also generated by thermal noise. The model is shown in Figure 3.26. The input-referred noise voltage $V_{n,n+p}$ due to thermal noise can be expressed as

$$V_{n,n+p} = \frac{4kT\text{BW}}{g_{m,n} + g_{m,p}}. \tag{3.33}$$

Where BW is the signal bandwidth. In a typical example assuming that the ratio of the channel width to the channel length of the nMOSFET and the pMOSFET is $(W/L)_{n,p} = 5, 10$, the amount of jitter is 25 fs for an input rising time of 50 ps.

COLUMN: If you do not use a S/H circuit
In a typical A/D converter, a S/H circuit is necessary to hold the input value during the A/D conversion. If the signal is constant or changing so slowly that the signal change during the conversion is less than 1 LSB, the S/H circuit is not necessary. Let us estimate the upper limit on the input frequency by assuming that this is the case.

If the input wave is a sinusoidal wave represented as

$$v(t) = q \left(\frac{2^N}{2}\right) \sin\left(2\pi f t\right), \tag{3.34}$$

its time derivative is

$$\frac{dv(t)}{dt} = q \left(\frac{2^N}{2}\right) 2\pi f \cos\left(2\pi f t\right). \tag{3.35}$$

Here, q is a voltage equivalent to 1 LSB. Therefore, for the signal change to be less than q during the time Δt required for the A/D conversion,

$$f \leq \frac{\frac{q}{\Delta t}}{q 2^N \pi} = \frac{1}{2^N \pi \Delta t} \tag{3.36}$$

must be satisfied. As an example, consider an A/D conversion with $N = 12$ and $f_s = 100$ kS/s. Under the assumption that the sampling period consists of the time required for A/D conversion Δt and the time required to store the digital output value in the external register. Assuming that the former is 8 μs and the latter is 2 μs, the upper limit on the frequency is found to be $f \leq$ 9.7 Hz, which is very low compared to the signal usually handled. However, if the output changes slowly like that from a temperature sensor, it might be unnecessary to use an S/H circuit.

The above discussion assumes that it takes some time Δt to quantize the input signal to the final bit resolution. This is the case for many A/D converters. However, as described later (Section 5.2), the flash A/D converter can in principle convert a high-speed input signal without using a S/H circuit. The reason for this is to quantize the analog input signal to the final resolution with a single sampling event.

3.2 Comparators

In A/D conversion, quantization is another important function besides sampling and is performed by a circuit called a comparator or quantizer[5]. As shown in Figure 3.27, the comparator has two analog inputs and one digital output. It compares the magnitude of the analog input V_{in} with the reference voltage V_{ref}, and outputs a digital value of 0 or 1 according to the comparison result. It is a type of amplifier that can momentarily amplify a small difference between analog inputs to the full-scale digital value. If 0 V and V_{DD} represent the logic values 0 and 1, respectively, the comparator output can be expressed as follows.

$$V_{out} = \begin{cases} V_{DD} & (V_{in} \geq V_{ref}) \\ 0 & (V_{in} < V_{ref}) \end{cases} \tag{3.37}$$

The comparator is essentially the same as a 1-bit A/D converter. It is a circuit operating just on the boundary between analog and digital, and the high-speed and high-resolution decision is required. In this section, circuits for realizing these functions are explained: first, comparators using an operational amplifier, and then multi-stage comparators, and latched comparators. After describing basic operation for each, technical challenges and solutions are described.

---

[5]The quantizer is also used when quantizing to 2 bits or more. On the other hand, the comparator is mainly used for 1-bit decision.

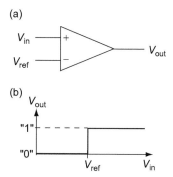

Figure 3.27 (a) Circuit symbol of a comparator and (b) input/output characteristics.

3.2.1 Opamp-based Comparators

3.2.1.1 Basic operation

First, consider an operational amplifier (opamp) shown in Figure 3.28. The output V_{out} is AV_{in} for a sufficiently small input V_{in}. Here, A is the gain of the opamp. If the input magnitude is somewhat large, the output is fixed at $+V_{\text{DD}}$ or $-V_{\text{DD}}$ as shown in Figure 3.28(b). The output voltage can thus be written as

$$V_{\text{out}} = \begin{cases} V_{\text{DD}} & (V_{\text{DD}}/A \leq V_{\text{in}}) \\ AV_{\text{in}} & (-V_{\text{DD}}/A \leq V_{\text{in}} < V_{\text{DD}}/A) \\ -V_{\text{DD}} & (V_{\text{in}} < -V_{\text{DD}}/A). \end{cases} \tag{3.38}$$

This is approximately the same as Equation (3.37) if $V_{\text{ref}} = 0$ V, except that the voltage representing 0 is $-V_{\text{DD}}$ instead of 0 V. Therefore, an operational amplifier with a high gain can be used as a comparator.

3.2.1.2 Clocked comparator

In A/D conversion, since a sampled value is quantized using a clock signal, a comparator that can be controlled by a clock is useful in many cases. An example is shown in Figure 3.29. ϕ_1 and ϕ_2 are two-phase non-overlapping clocks, and ϕ_{1a} goes LOW slightly earlier than ϕ_1. When ϕ_1 is HIGH, SW$_2$ and SW$_3$ are closed, and the circuit is in the reset mode. The opamp functions as a unity gain buffer. If the gain of the opamp is infinite, the output is 0 V. Also, both terminal voltages of capacitor C become 0 V, and the capacitor is completely discharged. When ϕ_2 is HIGH, the circuit is in the comparison mode. Since the inverting terminal of the opamp is high impedance, the charge stored in the capacitor does not change in the transition from the reset mode to the comparison mode. Thus, even after the transition, the charge is

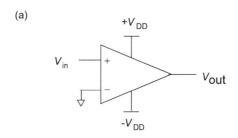

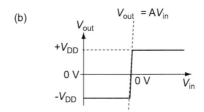

Figure 3.28   (a) Opamp-based comparator and (b) input/output characteristics.

zero so that the potential difference across the capacitor is also 0 V. Therefore, the voltage of the inverting terminal is equal to V_{in}, and the output V_{out} is determined by the sign of V_{in}.

A circuit for generating the non-overlapping clock is shown in Figure 3.30. If V_{clk} is LOW, ϕ_1 is HIGH and ϕ_2 is LOW. When V_{clk} changes from LOW to HIGH, ϕ_1 changes to LOW, this change is transmitted to the lower NOR gate, and then ϕ_2 goes HIGH after a delay of two inverters. In this way, two-phase non-overlapping clock signals are obtained, where HIGH does not overlap. A clock signal changing earlier than ϕ_1 is obtained at the node denoted by ϕ_{1a}.

The bent electrode in the capacitor symbol drawn in Figure 3.29(a) represents a lower or bottom electrode. Unless otherwise noted, this symbol is used in this book. A capacitor fabricated in an integrated circuit has the structure shown in Figure 3.31, and the electrode on the substrate side is called a lower or bottom electrode, and the other electrode is called an upper or top electrode. There is a relatively large parasitic capacitance between the lower electrode and the substrate. In contrast, the parasitic capacitance between the upper electrode and the substrate is small due to the electrostatic shielding effect of the lower electrode. For this reason, it is common to connect an upper electrode to a high-sensitivity terminal, such as an opamp input terminal, in which an even slight change in the potential is critical. The lower electrode

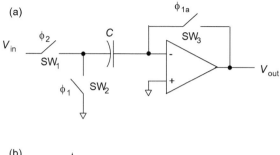

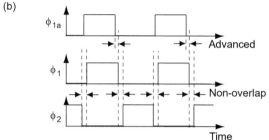

Figure 3.29   (a) Clocked comparator using an opamp and (b) timing chart.

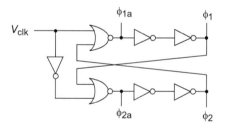

Figure 3.30   Non-overlapping clock signal generator.

is connected to a low impedance terminal that is not easily influenced by parasitic capacitance: for example, to the power supply or the opamp output.

Please note that capacitor C is not charged or discharged during the comparator operation shown in Figure 3.29. There is no need to charge the capacitor by the input signal source, which means that the input capacitance is small, reducing the power consumption[6]. If ϕ_1 and ϕ_2 are exchanged in

---

[6]Note, however, that the power consumption of opamp cannot be ignored. As will be explained in later chapters, reducing the power consumption of the opamp is often the main challenge in reducing the total power consumption of A/D converters using opamps.

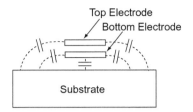

Figure 3.31   Typical structure of a capacitor used in integrated circuits.

Figure 3.29(a), charging and discharging occur for each comparison operation. Also, the decision result is not inverted. However, in this case, the output value is fixed in the comparison mode even if the input changes in the comparison mode. On the other hand, in the original circuit shown in Figure 3.29, when the input changes to cross 0 V in the comparison mode, the output is instantaneously inverted.

3.2.1.3 Charge injection, offset, and operation speed

As explained in Subsection 3.1.3 for the S/H circuit, the charge flows from the MOSFET channel to the capacitor, when the MOSFET turns off[7]. Since the comparator uses the MOSFET as a switch, the same phenomenon occurs, resulting in an incorrect comparison. In the circuit shown in Figure 3.29, such an influence is reduced by adjusting switching sequence: turning off ϕ_{1a} slightly earlier than ϕ_1. Since the capacitor terminal connected to the opamp is high impedance, no charges flow from SW_2 to the capacitor even when SW_2 turns off. Although it is necessary to consider charge injection from SW_3 to the capacitor, the injected charge is independent of the input signal because the source and drain voltages of SW_3 are fixed to ground.

Next, consider the case where the offset V_{os} exists in the comparator input. As shown in Figure 3.32, the offset can be modeled with an ideal opamp and a voltage source of V_{os} representing the offset voltage. In the reset mode, ϕ_1, $V_x = V_{out} = (A/(A + 1))V_{os} \approx V_{os}$), so the charge equal to CV_{os} is accumulated in the capacitor. When the circuit becomes the comparison mode, ϕ_2, $V_x = V_{in} + V_{os}$, which shows that the offset voltage can be canceled. This holds even for the comparator described later. In practical opamps and comparators using MOSFETs, there is always an offset of several mV, so this method of canceling the offset is quite useful.

---

[7]More precisely, it is necessary to consider the charge accumulated not only in the channel but also the overlap area between the gate and source and between the gate and drain.

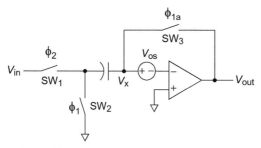

Figure 3.32 Clocked comparator using an opamp with offset.

A drawback of comparators using an opamp is that it is difficult to operate at high speed. For example, if we assume a comparator with an output digital voltage swing of 5 V and a resolution of 0.5 mV, the required gain A_0 (in dB) is

$$A_0 = 20 \log \frac{5}{0.5 \times 10^{-3}} = 80 \, \text{dB}. \tag{3.39}$$

If the input full scale is 5 V, the resolution is around 13 bits ($= \log_{10} (5/0.0005)/\log_{10} 2$). The transfer function of such an opamp is shown in Figure 3.33. If a unity-gain frequency f_T[8] of 10 MHz is assumed, the 3-dB frequency $f_{-3\text{dB}}$ is

$$f_{-3\text{dB}} = \frac{f_T}{A_0} = 1 \, \text{kHz}, \tag{3.40}$$

and the corresponding time constant τ is obtained as

$$\tau = \frac{1}{2\pi f_{-3\text{dB}}} \simeq 0.16 \, \text{ms}. \tag{3.41}$$

Therefore, the application is limited to a low-frequency region. Note that if an opamp with a smaller gain, $f_{-3\text{dB}}$ is higher and operation speed would be improved at the cost of a reduction in the bit-resolution. Of course, another solution is to increase f_T by using advanced CMOS technology.

3.2.1.4 Useful configurations

Figure 3.34 shows a fully differential comparator that uses differential signals as the input and output. By adopting such a differential configuration,

---

[8]The frequency at which the gain of an amplifier becomes 0 dB.

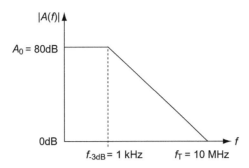

Figure 3.33   Open-loop opamp gain.

the influence of noise in the common mode signal can be efficiently suppressed. In many cases, the charge injection takes place in both the positive and negative channels simultaneously so that the influence on the common mode component is significant. The differential configuration can thus effectively suppress the influence of the charge injection. In recent years, differential comparators have been used in many data converters.

So far, comparators based on operational amplifiers were explained. Depending on required specifications, it is also possible to replace the opamp with a simple CMOS inverter. An example is shown in Figure 3.35. The operation of this circuit is almost the same as shown in Figure 3.29(a), but in the reset mode ϕ_1, V_x is the logical threshold voltage of the CMOS inverter. Here, the logical threshold voltage is an input voltage when the input and output voltages are the same. As explained in Figure 3.32, the difference

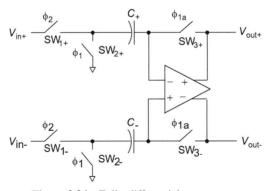

Figure 3.34   Fully-differential comparator.

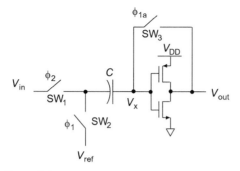

Figure 3.35 Comparator using a CMOS inverter.

between the logical threshold voltage and V_{ref} can be dealt with as offset voltage, which can be canceled. Therefore, if inverted, an output satisfying Equation (3.37) is obtained.

3.2.2 Multi-stage Comparators

A method for improving the operation speed of the comparator is to connect amplifiers with a small gain to form a multiple-stage comparator. An example is shown in Figure 3.36(a). Assuming that the gain of each stage is A_0, the delay time τ associated with each stage can be written as

$$\tau = \frac{A_0}{2\pi f_T}, \tag{3.42}$$

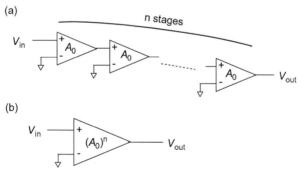

Figure 3.36 (a) Comparator consisting of n-stages of a small-gain (A_0) opamp and (b) single-stage comparator with a large-gain $((A_0)^n)$ opamp.

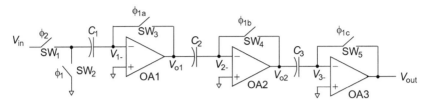

Figure 3.37   Offset cancellation in a multi-stage comparator.

where f_T is the unity-gain frequency of each amplifier. The total delay time τ_{total}, when connecting n opamps in series, can be approximated by

$$\tau_{\text{total}} \approx n\tau = \frac{nA_0}{2\pi f_T}. \tag{3.43}$$

On the other hand, the delay time τ_1 of a one-stage comparator using a single opamp with the same gain as shown in Figure 3.36(b) can be written as

$$\tau_1 = \frac{A_0{}^n}{2\pi f_T}. \tag{3.44}$$

The same value of f_T is used by assuming the same CMOS technology. Since $nA_0 \ll A_0{}^n$, it can be seen that a significant increase in speed is attained in the multi-stage comparator compared with the single-stage counterpart.

As mentioned in Subsection 3.2.1, non-ideal factors, such as the offset and charge injection, can be canceled by modifying circuit configuration as shown in Figure 3.37. The clocks denoted by ϕ_{1a} to ϕ_{1c} turn off in that order. ϕ_1 is turned off after all these clocks are turned off. As was explained in Subsection 3.2.1, this operation can cancel the offset of OA1. Besides, the charge injection from SW_2 is suppressed. Charge injection from SW_3 and offset of OA2 and OA3 are described below.

When ϕ_{1a} goes into the off state, charge injection from SW_3 causes V_{1-} to fall slightly because electrons flow out of SW_3. The voltage change in V_{1-} is multiplied by the gain of OA1 and transferred to the output V_{o1}. At this time, since ϕ_{1b} is still in the on state, the charge corresponding to this voltage is stored in C_2 together with the offset of OA2, and they are canceled in the comparison mode. Likewise, charge injection from SW_4 to C_2 is canceled. As a result, only the influence of charge injection from SW_5 to C_3 remains, but if it is referred to the input, it is equal to the change in V_{3-} divided by the product of the gains of OA1 and OA2. This can be almost ignored. As a disadvantage of this circuit, a clock generator to obtain several clock signals is necessary, which makes timing design somewhat complicated.

3.2.3 Latched Comparators

3.2.3.1 Basic circuit

Since the output of the comparator is a digital value, the linear amplification is not necessary. Therefore, instead of using an opamp, comparators using positive feedback to amplify an input voltage quickly to a digital value are widely used. Since they are similar to latch circuits used in digital ICs, they are called latched comparators.

An example of fully-differential latched comparators is shown in Figure 3.38 [46]. The differential configuration is generally suitable for high-speed operation. When V_{clk} is LOW, the circuit is in the reset mode, M_5 and M_6 turn on, and the output $V_{out\pm}$ becomes V_{DD}. Also, M_9 and M_{10} turn off, V_{x-} and V_{x+} are separated from V_{y-} and V_{y+}, respectively. When V_{clk} is HIGH, the circuit is in the comparison mode, M_5 and M_6 turn off, and the output $V_{out\pm}$ is separated from V_{DD}. Also, M_9 and M_{10} turn on, so that M_1 and M_3, and M_2 and M_4 constitute CMOS inverters. Since the output of one inverter is connected to the input of the other, if one output is 0 V, then the other output is V_{DD} in the stable state. If $V_{in+} > V_{in-}$, $V_{x+} > V_{x-}$. Therefore, the input of the inverter consisting of M_1 and M_3 is slightly smaller than the input of the other inverter consisting of M_2 and M_4 at the beginning of the comparison mode. This small difference is amplified over time, and V_{out+} becomes V_{DD} and V_{out-} becomes 0 V. Conversely, if $V_{in+} < V_{in-}$, V_{out+} is 0 V and V_{out-} is V_{DD}.

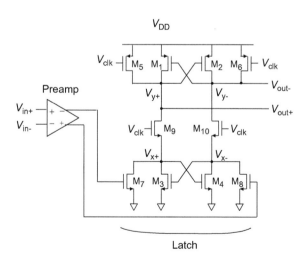

Figure 3.38 Differential latched comparator.

As shown in Figure 3.38, a preamplifier is often used to increase the resolution by amplifying the input signal in advance. However, note that if its gain is too large, the band narrows as was mentioned in Subsection 3.2.1. Typically, it is around 2 to 10 in many cases. Also, in comparators, a comparison result can remain in the circuit, which may affect the subsequent comparing operation. This is called a memory effect, or hysteresis. In the circuit of Figure 3.38, since the output is forced to V_{DD} in the reset mode, malfunction due to the memory effect can be avoided.

3.2.3.2 Dynamic comparator

When the output value is settled to 0 V or V_{DD} at the end of the comparison mode, the current path from V_{DD} to ground is blocked so that no static currents flow. Also in the reset mode, the current paths are blocked by M_9 and M_{10}, so no static power dissipation exists as is the case in conventional CMOS logic circuits. However, a constant current flows in the preamplifier, resulting in non-zero static power consumption.

As a solution to that problem, a dynamic comparator has been proposed as shown in Figure 3.39 [47]. As with CMOS dynamic logic circuits, this preamp operates using parasitic capacitance $C_{\pm}$. When V_{clk} is LOW, the preamplifier is in the precharge mode, $C_{\pm}$ is charged with on-state M_9 and M_{10} so that $V_{x\pm}$ is equal to V_{DD}. When V_{clk} is HIGH, the circuit enters the dynamic amplification mode, and the charge stored in $C_{\pm}$ is extracted as the drain current flowing through M_{13}. At this time, since the charge of $C_{\pm}$ varies depending on the magnitude of $V_{in\pm}$, a slight difference exists

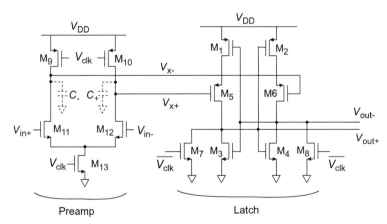

Figure 3.39 Dynamic latched comparator.

between $V_{x\pm}$. When the latch mode switches from reset to comparison, it amplifies the difference to obtain the comparison result as a digital value. In this preamplifier, there is no static current flowing from V_{DD} to ground. Therefore, there is no static power consumption, and low power consumption can be achieved.

The comparator output at the end of the comparison mode is a full-scale digital value, which can change at each comparison operation. If there is capacitive coupling between the output and the input of the comparator caused by MOSFETs and wiring, high-frequency components in the abrupt change in the output can leak into the input through the capacitive path. This is called kickback noise. Since a small analog signal is applied as the input, the influence may be substantial. Inserting a preamplifier can reduce the parasitic capacitance between the input and output, and then the kickback noise is suppressed. Also, a multi-stage configuration is useful for suppressing the kickback noise. Furthermore, when applying a clock signal to the tail current source M_{13} in Figure 3.39, the source potentials of M_{11} and M_{12} rapidly change. This is called clock feedthrough. Therefore, it is necessary to pay attention because the clock signal can leak to the input via the gate-source capacitance [48].

3.2.3.3 Metastability

Now, in order to analyze the operation speed of the comparator, the comparator just after entering the comparison mode is modeled by a simple circuit shown in Figure 3.38(a). The numbers assigned to the MOSFETs are the same as those in Figure 3.38. Figure 3.40(b) shows a small-signal equivalent circuit, where the MOSFET consists of a current source, an output resistor, and a gate capacitor. The node equations for V_x and V_y can be written as

$$\frac{A_v}{R_L}V_y = -C_L\frac{dV_x}{dt} - \frac{V_x}{R_L} \tag{3.45}$$

$$\frac{A_v}{R_L}V_x = -C_L\frac{dV_y}{dt} - \frac{V_y}{R_L}. \tag{3.46}$$

For simplicity, it is assumed that the transconductance, the output resistance and the gate capacitance of all MOSFETs are the same. Also, R_L is the combined resistance of R_{DS1} through R_{DS4}, whereas C_L is the combined capacitance consisting of C_{GS1} through C_{GS4}. Furthermore, since the voltage gain A_v is equal to G_mR_L, G_m is replaced with A_v/R_L.

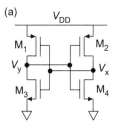

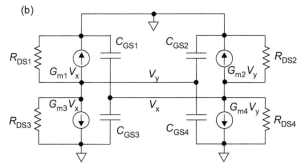

Figure 3.40   (a) Simplified circuit model of a comparator and (b) small-signal equivalent circuit.

Let us solve the simultaneous Equation (3.46). By subtracting each side and rearranging the terms,

$$\frac{\tau}{A_{\mathrm{v}} - 1} \frac{d\Delta V}{dt} = \Delta V \tag{3.47}$$

is obtained. Here, $\Delta V = V_{\mathrm{x}} - V_{\mathrm{y}}$ and $\tau = R_{\mathrm{L}} C_{\mathrm{L}}$. Solving this differential equation yields

$$\Delta V = \Delta V_0 \exp \frac{(A_{\mathrm{v}} - 1)t}{\tau} = \Delta V_0 \exp(t/\tau_{\mathrm{latch}}). \tag{3.48}$$

Here, ΔV_0 is the initial difference between V_{x} and V_{y} when the latch mode starts, and, τ_{latch} is an index of latch speed written as

$$\tau_{\mathrm{latch}} = \frac{\tau}{A_{\mathrm{v}} - 1} \simeq \frac{R_{\mathrm{L}} C_{\mathrm{L}}}{A_{\mathrm{v}}} = \frac{C_{\mathrm{L}}}{G_{\mathrm{m}}}. \tag{3.49}$$

If the digital swing is denoted as ΔV_{logic}, the time required for the comparator to amplify the input ΔV_0 to ΔV_{logic} is written as

$$T_{\text{latch}} = \frac{C_{\text{L}}}{G_{\text{m}}} \ln \frac{\Delta V_{\text{logic}}}{\Delta V_0}. \tag{3.50}$$

This equation indicates that when the initial input value ΔV_0 is small, T_{latch} can be large. In the extreme case, it takes quite a long time for the comparator to make a decision. This phenomenon is called metastability. Amplifying the input with a preamplifier will prevent metastability from occurring and helps speed up. However, note that even if ΔV_0 is somewhat large when the surrounding noise acts to reduce it, the metastability may take place. $G_{\text{m}}/C_{\text{L}}$ corresponds to the unity-gain frequency of the inverter. Comparing this with Equation (3.43), one can see that the latched comparator operates faster than the opamp-based comparator.

4

Digital/Analog (D/A) Converters

In this chapter, digital-to-analog converters (D/A converters or DACs) are described, while analog-to-digital converters (A/D converters or ADCs) will be dealt with in the next chapters. There is a reason for this order; since many A/D converters have built-in D/A converters, as was explained in Section 1.2, knowledge on D/A converters helps to understand A/D converters. This chapter starts by explaining the basic operation and performance specifications of D/A converters. Next, D/A converters using resistors (RDACs) and capacitors (CDACs) are explained. Note that several concepts introduced in the RDAC section are commonly applied to implementations of not only other D/A converters but also A/D converters. Finally, current-steering D/A converters (IDACs), which are suitable for high-speed operation, are described.

4.1 Basic Operation

The D/A converter generates an analog value corresponding to the input digital word. Figure 4.1 shows a typical setup for digital-to-analog conversion, consisting of a D/A converter and a reconstruction filter. The D/A converter converts the N-bit digital input D_{in} into the analog output V_{out}. V_{clk} and V_{ref} are the clock signal and the reference voltage, respectively. Figure 4.2 shows the waveforms of the clock and output voltages. It is assumed that the output value is updated at each positive edge (rising edge) of the clock signal. As shown in Figure 4.1, the reconstruction filter with low-pass characteristics smooths the stepwise output signal V_{out} into the analog signal $V_{out,f}$.

81

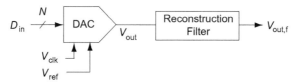

Figure 4.1   D/A converter followed by a reconstruction filter.

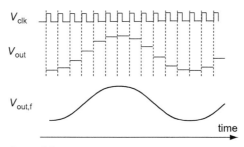

Figure 4.2   Output waveform of a D/A converter.

Usually, binary codes are used for digital input words. The output voltage V_{out} corresponding to the digital input $D_1 D_2 \cdots D_N$ can be expressed as

$$V_{\text{out}} = V_{\text{ref}} \left(D_1 2^{-1} + D_2 2^{-2} + \cdots + D_N 2^{-N} \right). \tag{4.1}$$

D_1 and D_N are the most significant bit (MSB) and the least significant bit (LSB), respectively. When the digital input changes from $000 \cdots 0$ to $111 \cdots 1$, the output voltage changes from 0 V to $V_{\text{ref}}(1 - 2^{-N})$. The voltage V_{LSB} defined as

$$V_{\text{LSB}} = V_{\text{ref}}/2^N. \tag{4.2}$$

represents the minimum voltage step in V_{out}. Also, 1 LSB is defined as

$$1\,\text{LSB} = 1/2^N. \tag{4.3}$$

Figure 4.3 shows the ideal input/output characteristics of a 3-bit D/A converter. For simplicity, V_{ref} is set to 1.

When the digital input changes, the actual D/A converter may generate spikes called glitches, as shown in Figure 4.4(b). Glitches are attributed to nonideal transitions in the digital input. Consider, for example, when the input changes from 0111 to 1000. If the MSB changes slightly earlier than the other bits, the input temporarily becomes 1111 in the middle of the transition, which generates the maximum value. On the contrary, if the change in the

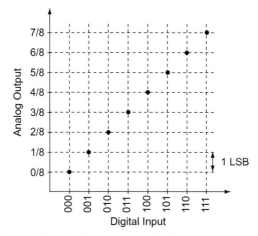

Figure 4.3 I/O characteristics of an ideal D/A converter.

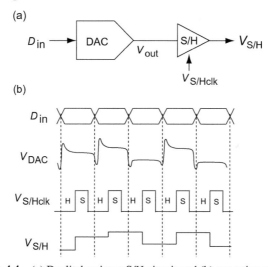

Figure 4.4 (a) Deglitch using a S/H circuit and (b) operation example.

MSB is slightly delayed compared with the other bits, the output settles to the final value 1000 via 0000 as the output of the intermediate state[1]. A S/H circuit shown in Figure 4.4 can remove glitches (called deglitch). V_{out} is sampled after it settles to the final value determined by Equation (4.1). Then setting the S/H circuit to the hold mode before the glitch generation can remove glitches in the output waveform.

---

[1]The circuit to adjust the timing is described in Section 4.5.

4.2 Performance Specifications

The input/output characteristics of an actual D/A converter often deviates from the ideal ones shown in Figure 4.3. The deviation is caused by several nonideal factors, such as process variations, temperature variations during operation, and parasitic elements. In this section, performance specifications representing static and dynamic characteristics are explained. The static characteristics are the relationship between the input and output when the input signal changes so slow that the transient response of the D/A converter is negligible. On the other hand, the dynamic characteristics represent the behavior when the operating frequency is high enough that the influence of the parasitic capacitance associated with circuit elements and interconnections cannot be ignored. In recent years, D/A converters operating in a high-frequency region have been widely used in communication and measurement applications, and the importance of dynamic characteristics is increasing.

4.2.1 Static Characteristics

In an ideal D/A converter, the output linearly increases as the input increases (Figure 4.3). However, in reality, there exist deviations from an ideal straight line, and these are referred to as nonlinear errors. One of the measures representing the nonlinearity is differential nonlinearity (DNL) shown in Figure 4.5. For simplicity, $V_{\mathrm{ref}} = 1$ is assumed. In the ideal D/A converter, the change in V_{DAC} corresponding to 1 LSB, or V_{LSB}, is determined by Equation (4.2). The DNL_i corresponding to the i-th input is defined as the difference between the actual step height and the ideal one (V_{LSB}):

$$\mathrm{DNL}_i = V_{\mathrm{DAC},i} - V_{\mathrm{DAC},i-1} - V_{\mathrm{LSB}}. \qquad (4.4)$$

Here, $V_{\mathrm{DAC},i}$ represents the output value for the i-th input. The DNL is expressed in LSB units in many cases, and Figure 4.5(b) shows the DNL corresponding to the characteristics shown in Figure 4.5(a).

The integral nonlinearity (INL) is an alternative measure to describe nonlinearity. As shown in Figure 4.6, it is represented by the change from the ideal output corresponding to each input word. If the INL in the i-th input code is represented by INL_i,

$$\mathrm{INL}_i = \sum_{k=0}^{i} \mathrm{DNL}_k \qquad (4.5)$$

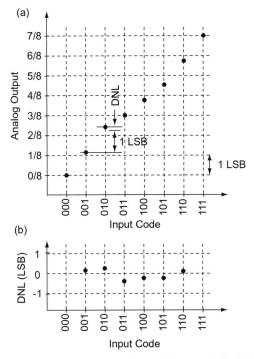

Figure 4.5 (a) D/A converter input/output characteristics and (b) differential nonlinearity (DNL).

holds. Here, it is assumed that the ideal output values are on the line that connects the start point with the endpoint.

Furthermore, a gain error and an offset error are used as performance specifications of the D/A converter. An example is shown in Figure 4.7. The gain error is the difference between the slopes of the line connecting the start point and the endpoint. There is another way to draw this straight line. For example, it may be drawn based on the least squares fitting. How to draw the line should be chosen according to each application. Sometimes the error in the vicinity of endpoints tends to be larger than that in the middle region. Also, considering the actual usage of the D/A converter, it is expected that the middle region is used more frequently region than the end regions.

4.2.2 Dynamic Characteristics

Figure 4.8 shows a typical spectrum of the output of a D/A converter with a sinusoidal input. Such a spectrum is used to evaluate dynamic characteristics.

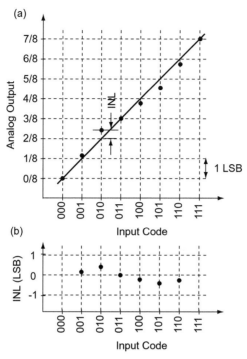

Figure 4.6 (a) D/A converter input/output characteristics and (b) integral nonlinearity (INL).

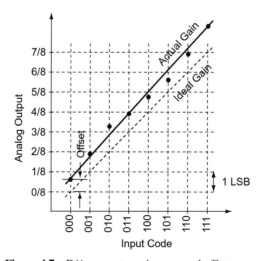

Figure 4.7 D/A converter gain error and offset error.

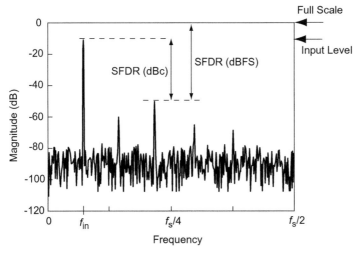

Figure 4.8 Spurious-free dynamic range (SFDR) of a D/A converter. $_{in}$ and $_s$ are the input frequency and sampling frequency, respectively.

One of the measures commonly used is the spurious-free dynamic range (SFDR) shown in this figure. If there is nonlinearity of the D/A converter, the output signal is distorted, and peaks appear at frequencies which are different from the input frequency f_{in}. The SFDR is defined by the difference between the input signal peak and the most significant peak representing the distortion. Since the distortion depends on the input frequency, it is necessary to specify the input and sampling frequencies in describing the SFDR.

The noise floor distributing around -90 dB in Figure 4.8 is due to quantization noise. The signal-to-noise ratio (SNR) can be obtained from the ratio of the signal power to the noise power. Also, the signal-to-noise-and-distortion ratio (SNDR) can be calculated as the ratio of the signal power to the sum of the harmonics power and the quantization noise power. These are the same as those used in analog circuits in general. Furthermore, as described in Section 2.2, the effective number of bits (ENOB) can be estimated from the SNR. The ENOB is smaller than the nominal number of bits determined from the circuit configuration because of various nonidealities included in actual circuits. Also, as the input frequency or the sampling frequency increases, these dynamic characteristics tend to deteriorate gradually. In particular, the input frequency at which the SNR is lower than that at low frequencies by 3 dB is called the signal-band frequency, or bandwidth of the D/A converter.

4.3 Resistor-based D/A Converters

There are two types of D/A converters using resistors: a voltage divider type and a current adder type. These are explained in this section.

4.3.1 Voltage Dividing

If M resistors with equal resistance values are connected in series, the reference voltage V_{ref} can be equally divided by M. A resistor string connected in series is called a resistor ladder. By selecting a node corresponding to the digital input, and then by connecting the node to the output terminal, the desired analog voltage can be obtained. As an example, a 3-bit D/A converter is shown in Figure 4.9. For a 3-bit input of $b_1 b_2 b_3$, the analog output V_{out} can be written as

$$V_{\text{out}} = V_{\text{ref}}(b_1 2^{-1} + b_2 2^{-2} + b_3 2^{-3}). \tag{4.6}$$

For example, if the input is 101, MOSFETs with gate voltages of b_1, $\overline{b_2}$, and b_3 turn on, and the output is equal to (5/8) V_{ref}. The output buffer in the figure is necessary to prevent the resistor-ladder current from flowing into the output terminal. If the current flows into the output, the current flowing through the resistors located above the node connected to the output becomes larger than the current flowing in the resistors located below it, and the output voltage decreases.

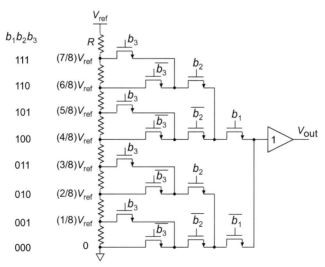

Figure 4.9   3-bit D/A converter using a resistor ladder.

The voltage at each node in the ladder gradually increases from the node connected to ground to the node connected to V_{ref}. Therefore, when the digital input increases, the output value is never reversed: *i.e.*, monotonicity is guaranteed. This is a significant feature of this type of D/A converters. On the other hand, if the resistance value varies, the accuracy in the output deteriorates. Such a mismatch of the resistance value depends on the manufacturing process, and typically the best accuracy obtained is around 10 bits. The laser trimming technique after manufacturing makes it possible to improve matching and to obtain a higher bit resolution. However, it would require an additional cost and time. Recently, instead of the laser trimming, the electronic calibration methods described below are widely introduced for high-resolution D/A converters.

A disadvantage of this type of D/A converters is its slow operating speed. Consider the signal delay from the node in the resistor ladder to the output node. The circuit between these nodes is an N-stage CR circuit shown in Figure 4.10. Here, a bit resolution is assumed to be N bits. R is the on-resistance of the MOSFET, C is the parasitic capacitance of the MOSFET and the interconnection. The signal delay time τ in this CR circuit can be expressed as[2]

$$\tau \approx RC \left(\frac{N^2}{2} \right) \tag{4.7}$$

In this example, $N = 3$. Equation (4.7) indicates that the delay time increases quadratically with an increase in N, and the operation speed becomes extremely slow as the bit resolution increases. At the same time, it is worth noting that the number of resistors in the ladder increases exponentially.

For high-speed operation, a decoder to switch only one MOSFET can be used as shown in Figure 4.11. The interconnection can be modeled as a

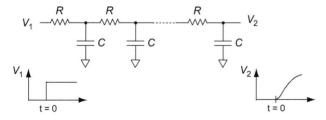

Figure 4.10   Signal delay in an RC circuit.

---

[2]It can be derived using the technique called zero-value time-constant analysis [30].

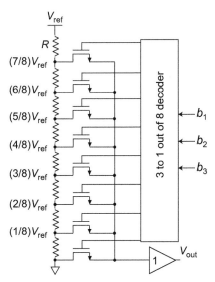

Figure 4.11   3-bit resistor-ladder D/A converter using an input decoder.

single-stage CR circuit, which reduces the delay time. However, because the source of the on-state MOSFET is connected to the source of all other off-state MOSFETs, the junction capacitance of these MOSFETs needs to be included in the delay time evaluation.

Another approach of interest is a two-step configuration shown in Figure 4.12. In this configuration, the exponential growth in the number of resistors can be efficiently suppressed. Here, the upper 3 bits are processed in the first stage, and the lower 3 bits are processed in the second stage so that 6 bits are obtained as the total resolution. The number of necessary resistors is $1/4 (= 2 \times 2^3/2^6)$ of Figure 4.9. As shown in this figure, if the input is 101011, the first stage nodes V_5 and V_6 are selected and used as the reference voltages in the second stage. Monotonicity is guaranteed if the unity gain buffers connecting the first stage and the second stage operates ideally. However, if there is a gain mismatch or offset, monotonicity is not guaranteed.

Return to Figure 4.9 and consider the cause of the glitch. Figure 4.13 shows a transition of the input from 100 to 011. As shown in Figure 4.9(a), if b_1 changes earlier than b_2 and b_3 by δt, the output changes to $100 \rightarrow 000 \rightarrow 011$, and 000 is generated as a temporary state. Conversely, if b_1 is delayed by δt, 111 is temporarily generated as an intermediate state. Using a S/H circuit as described in Section 4.1 effectively suppresses glitches.

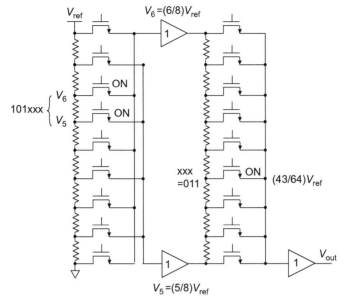

Figure 4.12 6-bit 2-step resistor-ladder D/A converter with a digital input of 101011.

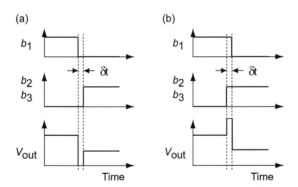

Figure 4.13 Glitch in D/A converter. When b_1 changed (a) earlier and (b) later than b_2 and b_3 by δt.

4.3.2 Current Adding

In the current-mode signal processing, addition can be performed by connecting wires. A D/A converter shown in Figure 4.14 converts the 4-bit input $b_1 b_2 b_3 b_4$ into the analog output by selecting necessary binary-weighted currents. The switch arrangement in this figure shows the case for 1001. Since $V_{\rm ref}/2R + V_{\rm ref}/16R$ flows through the feedback resistor R, the output

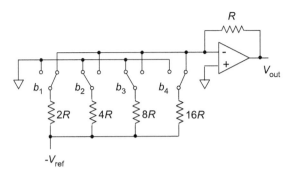

Figure 4.14   Current adder type D/A converter using resistors weighted by a factor of 2.

voltage becomes $(9/16)V_{ref}$, which agrees with the analog value corresponding to 1001. Note that if there is a mismatch in the binary-weighted resistors, monotonicity is not guaranteed. Also, note that there is a large difference in currents flowing through each switch by a factor of 2^N. It is necessary to scale MOSFET dimensions so that the all voltage drops across the MOSFET switches should be negligibly small. As described in the previous section, the glitch occurs if there are deviations between switching times.

If the bit resolution N increases, the circuit size shown in Figure 4.14 increases in proportion to N at first glance because adding one current path seems to be enough for obtaining one extra bit. However, it increases exponentially because the necessary resistance values increase exponentially, and then the occupied area also increases as well. As a solution to this problem, adding an attenuation resistor is proposed as shown in Figure 4.15. $3R$ in the figure is the attenuation resistor. However, because there remains a significant difference in currents flowing through the switches, the MOSFET scaling is necessary.

The R-2R ladder shown in Figure 4.16 is an extension of the concept of the attenuation resistor. When looking at the ladder from V_0, each branch has the same resistance value, and it is possible to divide the current by half. The D/A converter using the R-2R ladder is illustrated in Figure 4.17. It can be seen that the range of change in the resistance value is further reduced compared with that shown in Figure 4.15. However, note that the current difference is still substantial.

Another approach, where the monotonicity is the priority, is shown in Figure 4.18. This D/A converter uses a thermometer code instead of a binary code. As shown in Table 1.1, the thermometer code indicates the signal

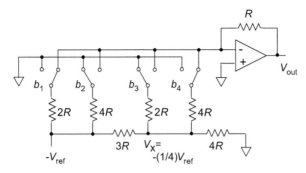

Figure 4.15 Current adder type D/A converter with an attenuation resistor.

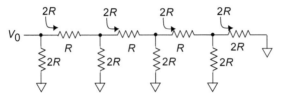

Figure 4.16 R-2R ladder

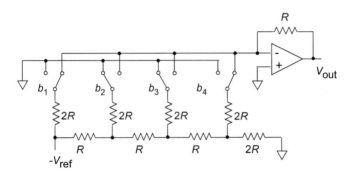

Figure 4.17 Current adder type D/A converter with an R-2R ladder.

magnitude by the number of 1's arranged in order from the lowest digit. Although the thermometer code is not the minimum representation of the numerical value, it guarantees the monotonicity. Also, it can suppress glitch generation. It seems that the number of resistors increases and the occupied area also increases. However, as shown in the example of capacitors with Figure 4.21, the occupied area is not increased so much because the entire resistors are constructed by combining a unit resistor. Since the current

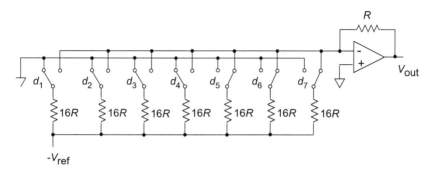

Figure 4.18   3-bit D/A converter with a thermometer code of 0000111 corresponding to a binary code of 011.

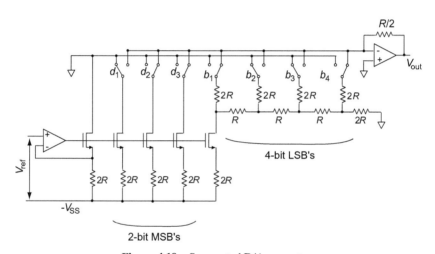

Figure 4.19   Segmented D/A converter.

values flowing through the MOSFETs are the same, scaling MOSFETs is not necessary for this circuit.

Utilizing these features, a resistance-based D/A converter using the thermometer code for the upper bits and the binary code for the lower bits has been proposed. It is called a segmented D/A converter. An example of a 6-bit D/A converter composed of the upper 2 bits and the lower 4 bits is shown in Figure 4.19. An R-2R ladder is used for the lower 4 bits to suppress an increase in the number of resistors. Also, a feedback loop is used to obtain $V_{ref}/(2R)$ as the current for the upper bit and the current for driving the R-2R ladder.

4.4 Capacitor-based D/A Converters

Many D/A converters using capacitors instead of resistors have been studied, mainly because if capacitors are used, the current flows only during charging and discharging, and no static currents flow. Therefore, capacitor-based D/A converters (CDACs) have been drawn attention in recent years because of its ability to operate with low power consumption. This section describes two types of D/A converters: voltage dividing type and charge sharing type. Besides, a serial type D/A converter consisting of only two capacitors is also proposed [49], which is described in Section 7.3.2.

4.4.1 Voltage Dividing

A voltage dividing 6-bit D/A converter is shown in Figure 4.20(a). This D/A converter operates as follows. First, all the switches from b_1 to b_6 are set to ground, and the reset switch is closed to discharge all the capacitors. Next, the reset switch is opened, and switches from b_1 to b_6 are connected to V_{ref} if the corresponding input is one. If the corresponding input is zero, the switches remain to be connected to ground. In this example, the input is assumed to be 010011, and an equivalent circuit of the capacitor array at this second phase is shown in Figure 4.20(b). As a result of voltage division by capacitors, $(19/64)V_{\text{ref}}$ is obtained. In general, the output voltage is represented as

$$V_{\text{out}} = \frac{\text{Sum of capacitance connected to } V_{\text{ref}}}{\text{Total capacitance}} \times V_{\text{ref}}. \qquad (4.8)$$

In order to change the input to another code and continue the D/A conversion, it is necessary to repeat from the reset operation.

The unity gain buffer is necessary to isolate the load capacitance at the output V_{out} from the capacitor array. In this connection, V_{x}, or the output

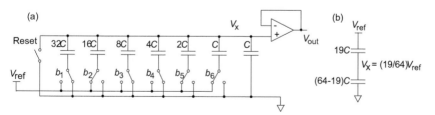

Figure 4.20 (a) 6-bit voltage dividing type D/A converter with an input code of 010011 and (b) equivalent circuit.

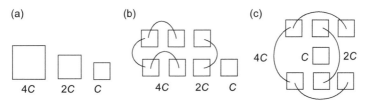

Figure 4.21   Capacitor layouts.

V_{out} can be any voltage values ranging from ground to V_{ref}. Therefore, good linearity is required for the unity gain buffer for a wide range of the input voltage.

Figure 4.21 shows layouts of binary-weighted capacitors. When the electrode area changes as shown in Figure 4.21(a), an accurate integer ratio cannot be obtained, because of the fringe capacitance[3]. Another cause of errors is attributed to dimensional errors occurring during pattern transfer. If unit capacitors are used as shown in Figure 4.21(b), even if there are errors in the absolute value of the capacitance, it becomes relatively easy to keep the integer ratio accurate. Furthermore, when placed as shown in Figure 4.21(c), it is possible to suppress the systematic process errors in element dimensions, such as the insulating film thickness between the capacitor electrodes. This technique is called a common-centroid layout, which is often used in analog circuit layout to match the characteristics of resistors and MOSFETs as well as capacitors.

As is the case in resistor-based D/A converters, to obtain a higher bit resolution, the necessary capacitance value increases exponentially, and the occupied area also increases[4]. A D/A converter using an attenuation capacitor as shown in Figure 4.22 [28] is proposed to avoid the exponential increase. The attenuation capacitor value C_{atten} is given by

$$C_{\text{atten}} = \frac{\text{Total capacitance for LSB}}{\text{Total capacitance for MSB}} \times C_{\text{unit}}, \qquad (4.9)$$

where C_{unit} is the unit capacitor value. When it is seen from V_{x}, the combined capacitance consisting of the attenuation capacitor and the capacitors on the left side is C. Therefore, a small capacitance can be effectively obtained

---

[3]This capacitance is due to the electric field leaking from the edge of electrodes, and proportional to the length of the perimeter.

[4]If the unit capacitance value is reduced, variations in characteristics would be significant, and the thermal noise mentioned in Section 3.1.6 would also increase. So, in practical designs, there is a minimum size for the unit capacitor.

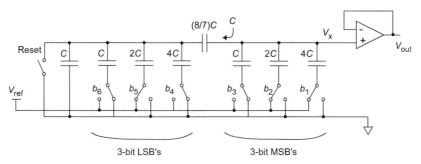

Figure 4.22 6-bit capacitor-based D/A converter using an attenuation capacitor with an input code of 010011.

without reducing actual capacitance sizes. However, as can be seen in this example, there is a problem that a non-integer capacitance ratio is required, which is not easy to achieve accurately.

4.4.2 Charge Sharing

A charge-sharing 6-bit D/A converter is shown in Figure 4.23. Figure 4.23(a) shows the precharge mode, where one of the capacitor terminals is connected to the inverting terminal of the opamp, which is virtually grounded. The other terminal of the capacitor is connected to V_{ref} if it corresponds to the digital input one, whereas it is connected to ground if it corresponds to the input zero. So, in this example, the input code is 010011. Figure 4.23(b) shows an equivalent circuit of (a). Figure 4.23(c) shows the charge-share mode. All capacitances are connected in parallel and put in the feedback path. As a result, as shown in Figure 4.23(d), the charge is shared by all the capacitors.

Since the upper electrodes are connected to the high-impedance inverting terminal, the total charge is kept constant in the transition from the precharge mode to the charge-share mode. Although there is parasitic capacitance associated with the lower electrodes, the effect can be negligible because the lower electrodes are connected to low impedance nodes both in Figure 4.23(a) and in Figure 4.23(b). Furthermore, by fixing the non-inverting terminal voltage to ground, the input common-mode voltage is fixed at ground. Therefore, compared with the voltage-dividing type described above, the requirement for the opamp design can be significantly relaxed. Also, similar to Figure 3.32, offset of the opamp can be canceled by setting the lower electrode of the capacitors to the ground before starting D/A conversion.

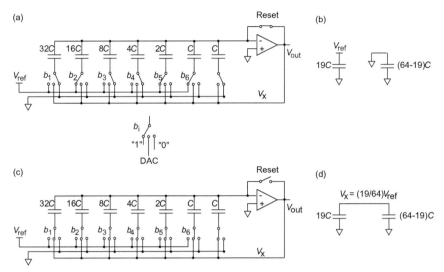

Figure 4.23 6-bit charge sharing type D/A converter. (a) Precharge mode and (c) charge-share mode. (b) and (d) represent corresponding equivalent circuits. The input code is 010011.

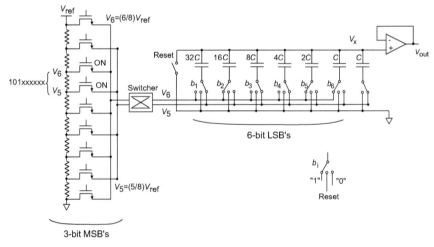

Figure 4.24 Hybrid type 9-bit D/A converter.

4.4.3 Hybrid

An example of a hybrid D/A converter that combines a D/A converter using capacitors and that using resistors is shown in Figure 4.24. According to the value of the upper 3 bits, the reference voltage used for the lower 6 bits is adjusted by the switcher.

In summary, in capacitance-based D/A converters, a reset operation is required at every sampling time. Also, there exists a settling time due to charging and discharging the capacitors. Therefore, they are suitable for low-power operation, but not suitable for high-speed operation, compared to resistance-based D/A converters.

4.5 Current-steering D/A Converters

A D/A converter which replaces the resistors used in the current adder type shown by Figure 4.14 with current sources is called a current-steering D/A converter. Since the high-speed performance is superior to the capacitor-based D/A converter, its applications to communication and measurement are drawing attention in recent years [50–52]. Figure 4.25(a) shows an example of a 4-bit current-steering D/A converter, and Figure 4.25(b) shows a circuit configuration used as a current source. High output impedance is obtained using cascode connection. V_{b1} and V_{b2} are the bias voltages to cascode transistors. As explained in Section 4.3, when the timing of the switch varies, glitches can occur. In this circuit, however, the glitch is suppressed by adjusting the timing of current steering using a D-type flip-flop. To obtain an accurate current ratio, it is necessary to use a layout technique similar to that for capacitors described in Figure 4.21. Eight unit MOSFETs connected in parallel are used in the current mirror for generating $8I_1$.

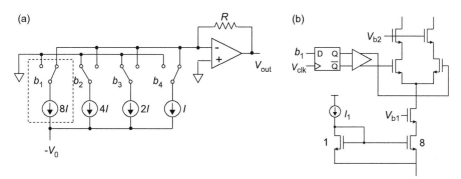

Figure 4.25 (a) 4-bit current-steering D/A converter and (b) current source shown by a dotted-line frame in (a).

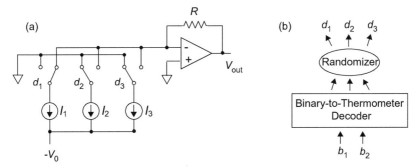

Figure 4.26   (a) Circuit diagram of a 2-bit current-steering D/A converter and (b) operation of dynamic element matching (DEM).

A thermometer-code-based D/A converter was already illustrated in Figure 4.18. Its advantage over binary-code-based D/A converters is that monotonicity is guaranteed. The same holds for current-steering D/A converters based on the thermometer code. Figure 4.26(a) shows a 2-bit D/A converter as an example. For 2-bit input codes of 00, 01, 10, and 11, corresponding thermometer codes, $d_1d_2d_3$, are equal to 000, 100, 110, and 111, respectively. Here, matching between the current sources, I_1 to I_3, is assumed. In practice, however, there is the mismatch between these, which results in nonlinearity in D/A conversion.

A calibration method to improve the matching has been proposed, which is shown in Figure 4.27 [53]. The switch position in the figure is for the calibration mode. The reference current $I_{ref} - I_0$ flows through the diode-connected M_1, and C_{gs} is charged so that the M_1 drain current is unchanged even after V_{cal2} is opened. When V_{cal1} are reversed, the circuit operates as a current source. The charge accumulated in C_{gs} determine the gate voltage of M_1, and the circuit operates as a current sink that draws the current equal to I_{ref}. Note that most of the current is realized with the current source I_0, and M_1 performs a fine adjustment. In order to apply this to the D/A converter shown by Figure 4.26(a), one extra current source is added to the three current sources. One of these operates in the calibration mode, and the others operate in the current sink mode. During operation of the D/A converter, matching

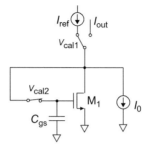

Figure 4.27 Current source calibration.

is improved by sequentially calibrating each current source one by one. The absolute value of each current source need not always be exactly I_{ref}. Instead, matching is the most critical.

If the same current sources are always selected in the D/A converter shown in Figure 4.26(a), a specific pattern due to the mismatch in the current sources is generated at the output, distorting the output signal. Another method to improve nonlinearity in the current-steering D/A converter is known as dynamic element matching (DEM) [54], which is shown in Figure 4.26(b). Here, $d_1 d_2 d_3$ is not a thermometer code but an m-of-3 code ($m = 0, 1, 2, 3$). For example, when the input is a binary code of 10, two out of d_1, d_2, and d_3 are selected to be 1, and the remaining one is 0, such that 110, 101, or 011. The selection is based on a certain algorithm, which will be explained in Section 6.5. The DEM is effective to suppress harmonic peaks, but it just spreads the harmonic components over a wide frequency range. Thus, the noise floor can slightly rise by using this technique.

Further advanced current-source calibration method [55] is shown in Figure 4.28 with a 3-bit D/A converter as an example. The current value from each current source is measured in advance. Suppose they are I_1 to I_7 as shown in Figure 4.28(a). According to the measurement result, they are combined as shown in Figure 4.28(b). Here, it is assumed that I_3 is closest to the average value. Figure 4.28(c) shows how to obtain eight levels required

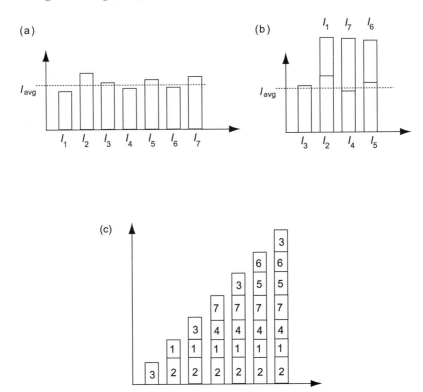

Figure 4.28   Improve matching by combining current sources. (a) Individual current source values, (b) combinations, and (c) realization of matching characteristics.

for 3-bit operation by using the combination. Linearity can be significantly improved compared with merely using I_1 to I_7 in this order. It is expected that similar kinds of digital-friendly methods will be further developed in the future.

5

Nyquist-rate Analog/Digital (A/D) Converters

A/D converters can be classified into Nyquist-rate A/D converters and oversampling A/D converters. The former operate at the Nyquist rate, or a sampling rate twice the signal bandwidth. In practice, it operates at a frequency slightly higher than twice. Nyquist-rate A/D converters are described in this chapter. The latter operate at a sampling rate of ten or more times the Nyquist rate. This will be described in Chapter 6.

A/D converters are utilized in diverse fields, and specifications required are also diverse. In order to meet a wide range of specifications, various types of architectures have been intensively studied. The A/D converter architectures that are widely used today are listed in Table 5.1. Figure 5.1 shows the range of operating speeds and resolutions covered by these A/D converters[1]. In this chapter, we will first describe the performance specifications of A/D converters. Then, the architecture and features of each A/D converter will be discussed, and various circuit examples are presented.

For readers who are interested in the first proposal of each architecture, please refer to the article describing the history of A/D converters [1, 21]. A/D converters based on the concept of artificial neural networks have also been reported [56, 57]; these are expected to create a new architecture that will incorporate both parallel processing and learning capability into A/D converters. However, since it is beyond the scope of this book, detailed descriptions are omitted.

---

[1]This shows typical areas covered by each type of A/D converters based on the current CMOS technology. Technological progress in the future can change boundaries.

103

Table 5.1   A/D converter architectures

Low/medium speed	Medium speed	High speed
High resolution	Medium resolution	Low/Medium resolution
Integral	Successive approximation (SAR)	Flash
–Section 5.7.1	–Section 5.4	–Section 5.2
Oversampling $\Delta\Sigma$	Algorithmic	Folding & interpolation
–Chapter 6	–Section 5.5	–Section 5.3
		Two-step –Section 5.5
		Pipelined –Section 5.6
		Time-interleaved –Section 5.8

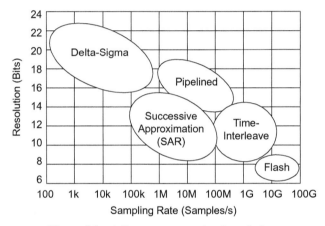

Figure 5.1   A/D converter speed and resolution.

5.1 Performance Specifications

Performance specifications of the A/D converter have much in common with those of the D/A converter described in Section 4.2. The difference is that the input is analog and the output is digital, which is the reverse of the D/A converter. In this section, we will focus on static characteristics that are different from D/A converters. As for dynamic characteristics, please refer to Section 4.2.

Input/output characteristics and quantization error of a 3-bit A/D converter are shown in Figure 5.2[2]. The quantization error V_Q is defined as the difference between the analog input V_{in} and the quantity V_{DAC} obtained

---

[2]This has already been shown as Figure 1.4. For the sake of convenience, it is shown here, again.

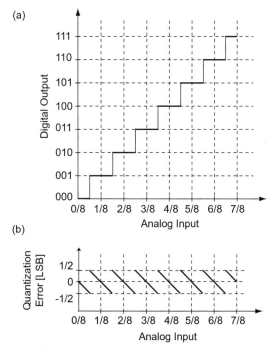

Figure 5.2 (a) Ideal A/D converter input/output characteristics and (b) quantization error.

by interpreting the digital output as a binary number

$$V_Q = V_{\text{DAC}} - V_{\text{in}}. \tag{5.1}$$

Here, V_{DAC} is expressed as

$$V_{\text{DAC}} = V_{\text{ref}} \left(D_1 2^{-1} + D_2 2^{-2} + \cdots + D_N 2^{-N} \right). \tag{5.2}$$

For simplicity, the reference voltage V_{ref} is set to 1 in Figure 5.2.

In practice, the digital output changes at the analog input value that are slightly different from the ideal value because of variations in device characteristics. Therefore, as shown in Figure 5.3(a), the step width changes from the ideal one (1 LSB). This amount of change is called a differential nonlinearity error (DNL). As shown in Figure 5.3(b), the DNL is expressed in LSB units. Another measure to express the nonlinearity is an integral nonlinearity error (INL). As shown in Figure 5.4(a), the INL is the deviation of the midpoint of the step and expressed in LSB units as well. As the DNL or INL becomes large, as shown in Figure 5.5, a specific code, for example, 101, may disappear. This is called a missing code.

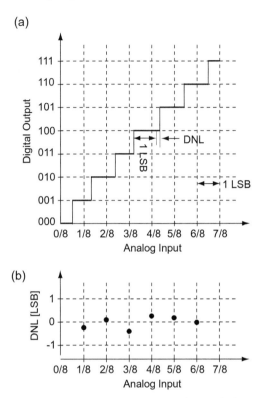

Figure 5.3 (a) Actual A/D converter input/output characteristics and (b) differential nonlinearity (DNL).

Figure 5.6 indicates gain and offset errors of a 3-bit A/D converter. This corresponds to those of the D/A converter indicated by Figure 4.7.

5.2 Flash A/D Converters

In the following, A/D converters listed in Table 5.1 will be explained one by one. First, flash A/D converters are described because they are the basis for other A/D converters. As an example, a block diagram of a 3-bit flash A/D converter is shown in Figure 5.7. As shown in the figure, the flash A/D converter is composed of a resistor ladder which divides the reference voltage V_{ref}, comparators for comparing each divided voltage with the analog input voltage V_{in}, and an encoder converting the thermometer code d_k generated by comparators to the binary code b_i. Here, $k = 1, 2, \cdots, 7$

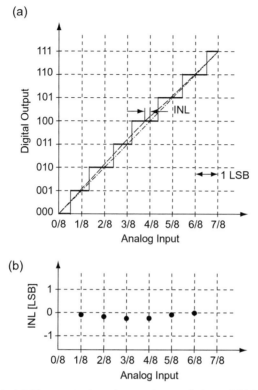

Figure 5.4 (a) Actual A/D converter input/output characteristics and (b) integral nonlinearity (INL). In (a), the broken line is a line connecting the midpoints of each step, and the dash-dotted line is one connecting the start point and the end point.

and $i = 1, 2, 3$ for the 3-bit case. The resistor ladder generates seven threshold voltages of $V_{ref}/16, 3V_{ref}/16, 5V_{ref}/16, \cdots , 13V_{ref}/16$. The comparators decide whether or not the input V_{in} is larger than these threshold voltages.

The latched comparator described in Section 3.2 is used, and each comparator is driven simultaneously by using the clock signal V_{clk}. If the clock- and input-signal delays can be negligibly small, the comparison is made precisely at the same time. Thus, in principle, the flash A/D converter does not need a S/H circuit. However, in practice, since the comparison times in each comparator might differ due to uneven delays in the signal and clock distribution network, distortions and bubble errors can take place as described below. Therefore, a S/H circuit is often inserted between the input and the comparator like other A/D converters.

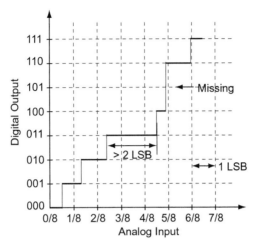

Figure 5.5   Missing code in an A/D converter.

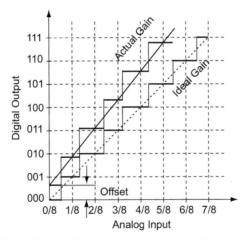

Figure 5.6   A/D converter gain error and offset error.

In order to obtain an N-bit resolution, it is necessary to prepare $2^N - 1$ threshold voltages to compare them with the input voltage. This means that $2^N - 1$ comparators are required[3]. As the bit resolution required increases, the number of comparators increases exponentially. Accordingly, the occupied area on a Si die, as well as power consumption, also increase exponentially. The upper limit on practical resolution of the flash A/D converter is then

---

[3]If overflow detection is required, 2^N comparators are needed.

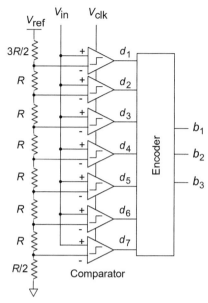

Figure 5.7 Block diagram of a 3-bit flash A/D converter.

about 8 bits or 10 bits at most. On the other hand, since many comparators are used in parallel,[4] the time required for A/D conversion is the shortest among all other conversion architectures, and it is possible to convert a high-speed signal exceeding 10 GHz.

A part of the encoder in Figure 5.7 is shown in Figure 5.8(a). Here, it is assumed that $3/16 < V_{in} < 5/16$. Thus, d_1 to d_5 are 0, and d_6 and d_7 are 1. Exclusive OR gates in the decoder detects the boundary between 0's and 1's in the thermometer code, and generates a 1-of-2^3 code, which is then converted to the binary code by a NOR-type ROM[5] in this example.

As already mentioned, the comparator output $d_1, d_2, \cdots, d_7$ is a thermometer code. However, the comparison result can be wrong because of uneven delay in signal distribution or comparator metastability and offset. The higher the sampling frequencies, the more likely an incorrect decision is made. For example, as shown in parentheses in Figure 5.8(a), zero and one may be interchanged between d_5 and d_6. As in this example, a 0('s) gets mixed in the part where 1's must follow is called a bubble.[6] When a bubble

---

[4]Therefore, it is sometimes called a parallel-comparison A/D converter.
[5]ROM: read-only memory.
[6]This is named after a similar phenomenon in a thermometer using mercury or alcohol.

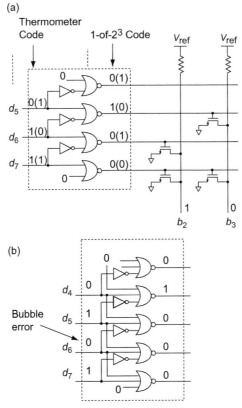

Figure 5.8   (a) Example of an encoder circuit and (b) bubble-error correction circuit.

error occurs, several 1's appear in the output of the comparator array, and no correct binary codes can be generated from the ROM. Figure 5.8(b) shows an example of an encoder circuit, which was proposed for correcting such a bubble error. As shown in this figure, even if the bubble error as shown above occurs, a proper 1-of-2^3 code is obtained, which makes it possible to prevent the reading operation of the ROM from failing.

The decoder shown in Figure 5.8(b) can deal with a simple bubble error where only one digit is wrong. However, bubble errors of more complicated patterns cannot be corrected. Figure 5.9 shows a Wallace-tree encoder, which has more enhanced correction capabilities. As shown in Figure 5.9(b), it consists of full adders (Figure 5.9(a)) constructing a tree structure. This encoder counts the total number of 1's in the comparator outputs and generates a binary code corresponding to a thermometer code in which the

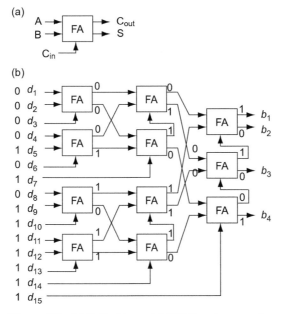

Figure 5.9 (a) Full adder and (b) Wallace tree encoder.

same number of 1's are sequentially arranged from the least significant digit to higher digits. In this figure, it is shown that for an input with a bubble errors in two digits, d_6 and d_8, the output binary code is 1001. The decimal equivalent is 9, which is equal to the number of 1's in the input.

Other output examples of a 3-bit Wallace-tree encoder are shown in Figure 5.10. For any bubble errors indicated by Y_1 to Y_4, the binary output obtained is equivalent to the thermometer code with the dotted-line boundary separating 0's and 1's. This agrees with the fact that the total number of 1's is four in each input. Using the Wallace tree can make the circuit larger. However, if state-of-the-art scaled-down technology is used, the increase in the occupied area and power consumption is not a severe problem. For this reason, in recent years, this type of encoders is frequently used in flash A/D converters.

As described above, in the flash type A/D converter, there is a limit in the bit resolution around 8 to 10 bits at most. However, it operates at the highest speed among other A/D converters. In addition, it has a small latency between the input and the output. Therefore, it is used in applications where high-speed real-time performance is the priority. For example, the flash A/D converter is widely used in ultra-high-speed optical fiber communication

$$Y_1 \; Y_2 \; Y_3 \; Y_4 \quad \rightarrow \quad D_{\text{Wallece}}$$

	Y_1	Y_2	Y_3	Y_4	D_{Wallece}
d_1	0	0	0	0	0
d_2	0	0	1	0	0
d_3	0	0	0	1	0
d_4	1	1	0	1	0
d_5	0	1	0	0	1
d_6	1	1	1	0	1
d_7	1	0	1	1	1
d_8	1	1	1	1	1

Figure 5.10   Output examples of the Wallace tree.

receivers and high-speed Ethernet receivers as well as in ultra-high-speed measurements such as multi-GHz digital oscilloscopes and radio astronomy. It has also been conventionally used in read channels in hard disk drives. As will be described in Section 5.8, recently, speeding up based on the time-interleave configuration has progressed, and in the future, it will take over flash A/D converters in some applications. However, the superiority of the flash A/D converter in the highest speed region is still expected to continue. It should also be pointed out that flash A/D converters are widely used as sub-A/D converters incorporated in other types of A/D converters.

Readers who want to learn further, refer to papers [58, 59] related to the flash A/D converter listed at the end of the book.

5.3 Folding and Interpolation A/D Converters

The flash A/D converter described in the previous section required 2^{N-1} comparators to obtain a resolution of N bits. A method capable of reducing the number of comparators while maintaining high-speed conversion characteristics is described in this section. In the flash A/D converter, the number of comparators, which are involved in the critical decision, is not so large. In other words, only a few comparators, with two inputs close to each other, determine the final digital output. Such an observation allows one to use comparators more efficiently.

To find out the method to realize it, consider the relationship between the decimal numbers from 0 to 15 and the corresponding binary codes as shown in Table 5.2. It can be seen that the lower 2 bits repeat the same pattern as the decimal number increases. This regularity leads an idea that the A/D conversion can be divided into two parts: One is for the upper 2 bits and

Table 5.2 Encoding from decimal to binary

Decimal	Upper 2 bits	Lower 2 bits	Lower 2 bits Folded	Lower 2 bits Cyclic Thermometer Code
0	00	00	00	00
1	00	01	01	01
2	00	10	10	11
3	00	11	11	10
4	01	00	11	00
5	01	01	10	01
6	01	10	01	11
7	01	11	00	10
8	10	00	00	00
9	10	01	01	01
10	10	10	10	11
11	10	11	11	10
12	11	00	11	00
13	11	01	10	01
14	11	10	01	11
15	11	11	00	10

the other one is for the lower 2 bits to obtain 4-bit resolution. The number of comparators is then $3 + 3 = 6$. Since conventional flash A/D converters need $2^4 - 1 = 15$ comparators for 4-bit resolution, significant reduction is possible.

The concept of a 6-bit folding ADC is shown in Figure 5.11. As shown in Figure 5.11(b), the folding circuit has the repeating characteristics corresponding to the lower bits of Table 5.2. The triangular-shaped characteristic shown in Figure 5.11(c) is more practical to implement than the saw-tooth ones shown in Figure 5.11(b). I/O characteristics of the 6-bit folding ADC are shown in Figure 5.12. For example, when the input is $\left(\frac{21}{64} <\right)\frac{43}{128}\left(< \frac{22}{64}\right)$ shown by the arrows, the upper 3 bits are 010, and the lower 3 bits are 101.

An example of folding circuits using BJTs is shown in Figure 5.13. Figure 5.13(a) illustrates an emitter-coupled differential pair used as a unit circuit. If $V_{in+} \ll V_{in-}$, I_0 flows through R_2, and $V_{out+} = V_L$ and $V_{out-} = V_H$. On the other hand, If $V_{in+} \gg V_{in-}$, I_0 flows through R_1, and $V_{out+} = V_H$ and $V_{out-} = V_L$. The resulting characteristics are shown in Figure 5.13(b). By combining four differential pairs as shown in Figure 5.13(c), the characteristics shown in Figure 5.13(d) can be obtained. It is also possible to replace BJTs with MOSFETs.

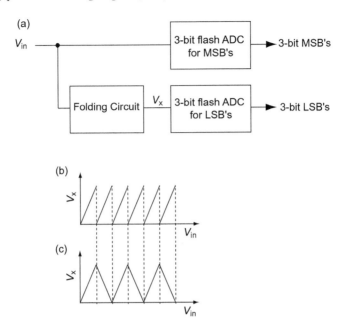

Figure 5.11   (a) Block diagram of a folding type ADC, (b) input/output characteristics necessary for a folding circuit, and (c) those realized by a practical circuit.

The problem with this circuit is shown in Figure 5.14(a). In principle, the bending point should be sharp as shown by the dotted line, but in reality, the corners are rounded as shown by the solid line, so conversion errors occur. Although it can be improved to some extent by optimizing the circuit parameters in Figure 5.13(c), still it is difficult to obtain perfectly sharp folding characteristics as shown in Figure 5.11(c). One solution for suppressing the error due to such nonlinearity is shown in Figure 5.14(b) [60]. In this case, conversion errors are eliminated by using two folding circuits. By selecting V_i' between V_i and V_{i+1}, and by using the straight part of one circuit for the curved part of the other circuit, it is possible to realize a folding characteristic with good linearity for any input values.

It is interesting to note that even one folding characteristic can halve the number of comparators. In this case, instead of using complex folding circuits, a single comparator is enough for the MSB decision, followed by a switch exchanging the differential inputs according to the MSB [61]. Thus, sharp bending characteristics can be obtained by using a comparator.

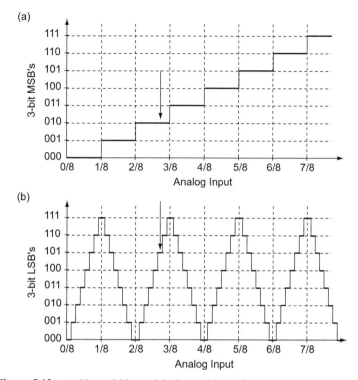

Figure 5.12 (a) Upper 3 bits and (b) lower 3 bits of a 6-bit folding type ADC.

Another way to solve the conversion error due to the rounding of the output of the folding circuit is based on the nature that the position of zero-crossing points in the transfer characteristics is independent of the rounding. The zero-crossing point is the point where the differential outputs become zero, *i.e.*, $V_{out+} = V_{out-}$. A block diagram of a 3-bit folding A/D converter is shown in Figure 5.15(a). The two folding circuits are designed such that the differential outputs V_1, V_2 are shown in Figure 5.15(b). The digital outputs are thus 000, 001, 011, 010, $\cdots$, which are cyclic thermometer codes corresponding to 0 to 7 in decimal shown in the rightmost column of Table 5.2.

A further extension of this concept is a folding-and-interpolating A/D converter [62]. An example is shown in Figure 5.16. As shown in Figure 5.16(a), V_{M+} and V_{M-} are generated by voltage divider using

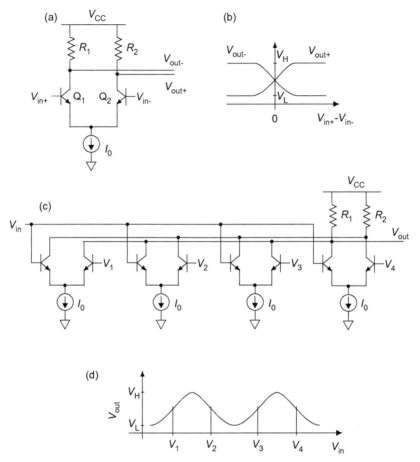

Figure 5.13 (a) BJT differential pair and (c) BJT folding circuit. (b) and (d) are corresponding input/output characteristics.

resistors. This method is called interpolation. Each waveform is plotted in Figure 5.15(b). Here,

$$V_1 = V_{1+} - V_{1-} \tag{5.3}$$

$$V_2 = V_{2+} - V_{2-} \tag{5.4}$$

$$V_{M+} = \frac{1}{2}(V_1 + V_2) \tag{5.5}$$

$$V_{M-} = \frac{1}{2}(-V_1 + V_2). \tag{5.6}$$

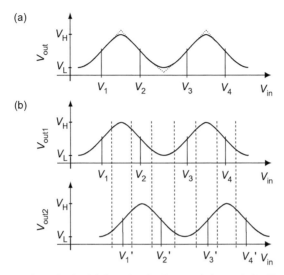

Figure 5.14 (a) Nonlinearity in folding circuit characteristics and (b) Combination of two characteristics to improve linearity.

Table 5.3 Cyclic thermometer code

Binary Code	Cyclic Thermometer Code
000	0000
001	0001
010	0011
011	0111
100	1111
101	1110
110	1100
111	1000

If 0's and 1's are determined by the signs of these voltages and the results are written from top to bottom, a 4-digit cyclic thermometer code shown in Table 5.3 is obtained. Actually, as the input increases, the output changes from 0000 to 0001, 0011, $\cdots$, and 1000. These eight codes correspond to a 3-bit resolution. In other words, by replacing the LSB part of the circuit shown in Figure 5.15(a) with the interpolation circuit shown in Figure 5.16, the resolution can be improved by 1 bit.

If the number of resistors used for the voltage division increases, the number of zero-crossing points also increases, which increases the number of cyclic thermometer codes. This means that the resolution can

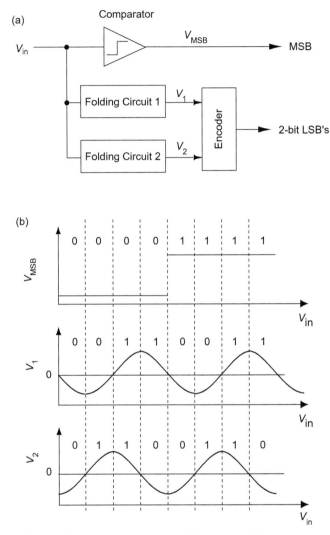

Figure 5.15   (a) Block diagram of a 3-bit folding type ADC and (b) its operation.

be further improved. However, considering the nonlinearity of interpolation characteristics, it is difficult to arrange the zero-crossing points accurately at equal intervals. Strictly speaking, only in the case of dividing into two as shown in Figure 5.16, they are arranged at equal intervals, unless there are mismatches in circuit element characteristics. As a precaution when using the folding circuit, be aware that the folding circuit needs to operate at high

(a)

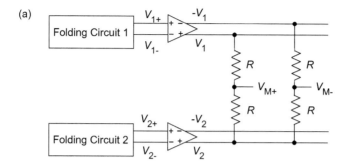

(b)

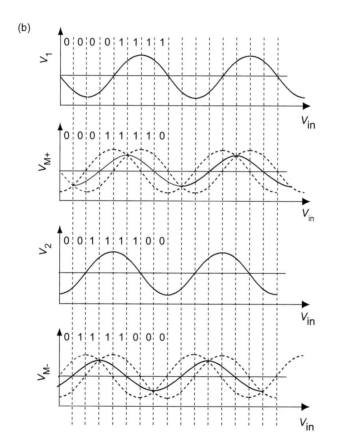

Figure 5.16 (a) Interpolation circuit and (b) output of a cyclic thermometer code.

speed. If a full-scale sinusoidal input of frequency f_{in} is assumed, the output of the folding circuit oscillates at a frequency of $f_{in} \times N_{fold}$. Here, N_{fold} represents the number of folding. In Figure 5.13, $N_{fold} = 4$.

As described above, in the folding and folding-and-interpolating A/D converters, it is possible to reduce the number of comparators and preamplifiers while keeping the same high speed as the flash A/D converter. Therefore, they have the advantage that it is possible to reduce power consumption and simplify the circuit. Meanwhile, a folding circuit and an interpolation circuit are necessary, and careful design is required for matching and high-speed performance.

5.4 Successive-approximation A/D Converters

5.4.1 Binary Search Algorithm

Consider a game in which one finds the number the other has chosen. The tactics for reaching the answer are as follows. If the number is chosen, for example, from integers between 0 and 255, one should first ask whether or not the number is larger than 127. If the answer is "yes," then the next question is whether it is larger than 191. If the answer is "no," then it is if it is larger than 63. The range can be narrowed by repeating such questions, and one can reach the right number chosen by the other. This is an example of a binary search algorithm.

Successive-approximation A/D converters utilize this algorithm [63]. Figure 5.17 shows an example of operation in the case of 3 bits. Here, it is assumed that the input is in the range from 0 to 1. First, it is decided whether or not the analog input value V_{in} is larger than 4/8 ($= 1/2$). In this example, since V_{in} is not larger than 1/2, the output for the most significant bit (MSB) is 0. Next question is whether it is larger than 2/8 ($= 1/4$). This time V_{in} is larger than 1/4, so the second most significant bit is 1. Finally, since it is larger than 3/8, the least significant bit (LSB) is 1. By putting them together, the A/D converter generates a digital output of 011. If this operation is repeated N times, in principle, an N-bit digital output is obtained.

Figure 5.18 shows a block diagram for executing the binary search algorithm mentioned above. The S/H circuit is used to hold the time-varying analog input V_{in} during the conversion from the most significant bit to the least significant bit. Unlike the flash type, the S/H circuit is indispensable. A reference signal for comparing with V_{in} is generated using an internal D/A converter (DAC). Depending on the comparison result, the reference

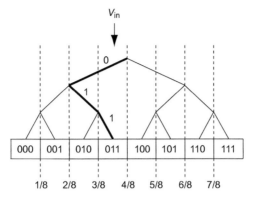

Figure 5.17 3-bit binary search tree.

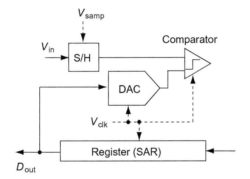

Figure 5.18 Block Diagram of a SAR Type A/D Converter.

voltage signal is sequentially updated. The register generates the digital input code to the D/A converter. The output of the D/A converter is an approximate value of V_{in}, which gradually approaches V_{in} one after another as the comparison cycle is repeated, so this is called successive approximation. The register, called successive-approximation registers (SAR), stores the comparison results to generate an N-bit digital word after N-cycles of repetition. This is the reason why it is called a SAR A/D converter.

A typical timing chart for operating a SAR A/D converter is shown in Figure 5.19. V_{in} is sampled by V_{samp}. The inverse of the period of V_{samp}, $1/T_s$, is the sampling frequency. Then, N pulses are used as the internal clock signal V_{clk} to obtain an N-bit digital output. If the sampling is performed for one cycle of V_{clk}, an internal clock frequency as high as $(N + 1)/T_s$ is required.

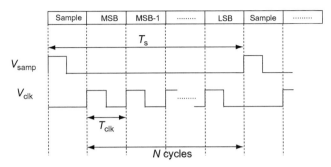

Figure 5.19   Timing chart of a SAR A/D converter.

Among D/A converters described in Chapter 4, a capacitor-based D/A converter (CDAC) is used as the DAC in Figure 5.18. This is because the capacitor array in the D/A converter can be used as the holding capacitor of the S/H circuit as will be described below. Another reason is that the static power consumption is small in the CDAC.

As mentioned above, an N-bit SAR A/D converter needs N cycles at least to complete the A/D conversion. Thus, it is inferior to the flash A/D converter concerning the operating speed. However, it has an excellent advantage that low-power consumption operation is possible because one comparator is enough, the static power consumption of CDAC can be small, and power-hungry operational amplifiers are unnecessary. In other words, designing SAR A/D converters based on miniaturized CMOS technology is relatively easy. Also, the capability of low-power operation is consistent with a recent trend of VLSI technology pursuing low-power consumption. It is often said that the SAR A/D converter is scaled-down friendly and is in the limelight, recently.

The comparator and CDAC limit the conversion speed in the SAR A/D converter. Specifically, it is necessary to consider the metastability of the comparator described in Section 3.2.3. Also, the response time of the D/A converter due to charging and discharging of capacitors cannot be neglected. Therefore, if the SAR operation is synchronized with an external clock V_{clk} with a constant frequency, its frequency should be designed by taking the worst case into account to guarantee the correct comparison. A disadvantage is that unnecessary waiting time can occur when a correct comparison operation is completed within a time considerably shorter than the external clock period. An alternative method has been proposed in which the SAR A/D converter itself judges the end of comparison and starts the next comparison cycle. This is called the asynchronous scheme [64] and has been drawing attention in recent years. Since not all judgments suffer from metastability, it is possible to shorten the total comparison time and speed up the SAR A/D converter.

5.4.2 Binary Search with Capacitor-based DAC

5.4.2.1 Charge redistribution

The binary search algorithm can be implemented by using the charge redistribution D/A converter[7]. To understand this, first, let us consider a 2-bit example shown in Figure 5.20. In Figure 5.20(a), no charges are initially stored in the two capacitors. If V_{ref} and 0 V are applied to the bottom

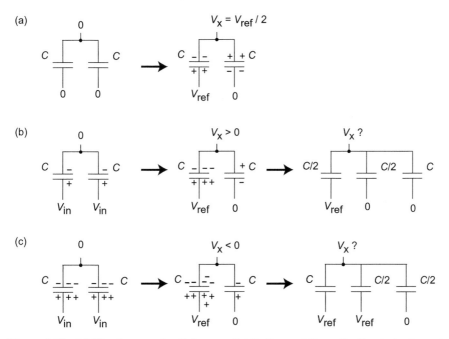

Figure 5.20 (a) Simple example of charge redistribution and its application to implement binary search algorithm when the input voltage V_{in} is (b) small ($V_{in} < V_{ref}/2$) and (c) large ($V_{in} > V_{ref}/2$).

---

[7]It is basically the same as the voltage-division type D/A converter described in Section 4.4, but this term is conventionally used in the field of A/D converter. The first proposed paper [65], and well-cited articles [66] are listed at the end of the book.

electrodes as shown in the figure, then $V_x = V_{ref}/2$ is satisfied. It is assumed that no external current flows in or out of the node V_x connecting the two capacitors.

Next, consider a situation where a small input voltage V_{in} is sampled as shown in Figure 5.20(b). After the input voltage is removed, V_{ref} and 0 V are applied to the two capacitors. In this case, a charge redistribution takes place in a similar manner described in Figure 5.20(a). Negative charges are induced in the upper electrode of the left capacitor, and as a result, positive charges are accumulated in the upper electrode of the right capacitor as shown in Figure 5.20(b). Therefore, $V_x > 0$. Intuitively, if V_{in} is small enough, V_x is considered unchanged from Figure 5.20(a), so $V_x > 0$.

On the other hand, if V_{in} is sufficiently large as shown in (c), a large number of negative charges are generated in the upper electrode of both capacitors. Even if V_{ref} is applied to the lower electrode of the left capacitor, the number of negative charges generated at the upper electrode of the left capacitor is not large enough to generate positive charges on the upper electrode of the right capacitor. Negative charges then remain in the upper electrode of the right capacitor, and $V_x < 0$.

Let us estimate the input voltage V_{in} that determines whether V_x is positive or negative. When V_{in} is applied, the total charge accumulated in the upper electrodes of the two capacitors is $-2CV_{in}$. Since there is no current flowing into or out of the node connecting the upper electrodes of the two capacitors, the sum of the charges accumulated in the upper electrodes is constant before and after the application of V_{ref} and 0 V. Then $-2CV_{in} = C(V_x - V_{ref}) + CV_x$ is hold, and $V_x = -V_{in} + V_{ref}/2$ is obtained. Therefore, if $V_{in} < V_{ref}/2$, $V_x > 0$, and if $V_{in} > V_{ref}/2$, $V_x < 0$. The first step in binary search algorithms can be implemented in this way.

Now consider the second step of the binary search. When $V_{in} < V_{ref}/2$, divide the capacitor connected to V_{ref} in half, and apply V_{ref} to one of them and 0 V to the other as shown in Figure 5.20(b). Then, the negative charges accumulated in the upper electrodes of these capacitors decreases. When calculated in the same manner as mentioned above, $V_{ref}/4$ is obtained as the value of V_{in} which determines the sign of V_x. On the other hand, when $V_{in} > V_{ref}/2$, the capacitor connected to ground is divided in half: One of which is applied with V_{ref}, and the other with 0 V. The negative charges accumulated in the upper electrodes decreases. The boundary that determines the sign of V_x is then obtained as $3V_{ref}/4$. By repeating these steps, the circuit can realize the binary search algorithm shown in Figure 5.17.

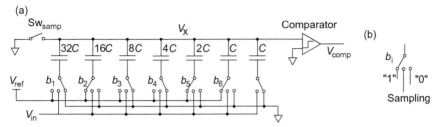

Figure 5.21 (a) Circuit diagram of a SAR A/D converter and (b) function of a switch terminal.

Based on the discussion above, a 6-bit SAR A/D converter can be constructed as shown in Figure 5.21. In the sampling mode, Sw_{samp} is closed, and all the capacitors are connected to V_{in}. At this time, all the capacitors are charged with V_{in}. Next, Sw_{samp} is opened, and all the capacitors are connected to ground. Then, $V_x = -V_{in}$, which means that the input voltage is sampled. The MSB can be obtained as the comparator output V_{comp} by switching only b_1 to "1" (or V_{ref}) and leaving other switches to "0". Since the capacitor $32C$ used for the MSB decision is equal to the sum of all other capacitors, the situation described in Figure 5.20 is realized. When MSB $= 1, b_1$ is kept at "1" and b_2 is switched to "1". When MSB $= 0$, b_2 is switched to "1" after b_1 is returned to "0". The second MSB is obtained as the output of the comparator V_{comp}.

By repeating these steps, a 6-bit output is obtained. If the switching is completed with the arrangement shown in Figure 5.21, the output is 010011. It should be noted that after the 6th cycle, V_x represents the quantization error, which can be utilized for pipelining SAR A/D converters or noise-shaping characteristics as will be explained in Section 7.3. As with the capacitor DAC of Figure 4.20, the final voltage V_x is

$$V_x = -V_{in} + \frac{C_T}{C_T + C_B} V_{ref}, \tag{5.7}$$

where C_T and C_B are the sums of the capacitances connected to V_{ref} and ground, respectively. V_x approaches 0 as the switching sequence progresses. For this reason, the influence of the common mode voltage dependence of the offset of the comparator can be small.

When the required bit resolution is increased, the total capacitance of the capacitor array increases exponentially. In order to prevent this, it is useful to use the attenuation capacitor shown in Figure 4.22. However, since the ratio of this capacitance value to the unit capacitance becomes a non-integer, good matching is difficult to achieve.

5.4.2.2 Charge sharing

A charge-sharing scheme [67] is known as another implementation for realizing the binary search algorithm. A 4-bit example is shown in Figure 5.22. The procedure is similar to weighing by using a balance with binary weights. First, the binary-weighted capacitor array is charged with V_{ref}, while the input is sampled by the sampling capacitor $16C$, as shown in Figure 5.22(a). Next, the largest capacitor in the array $(8C)$ is connected to the sampling capacitor as shown in Figure 5.22(b). By a simple calculation similar to one mentioned above,

$$V_{x1} = \frac{16}{24}\left(-V_{in} + \frac{1}{2}V_{ref}\right)$$

(5.8)

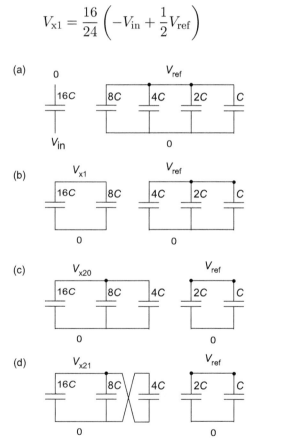

Figure 5.22   Binary search algorithm implemented by a charge sharing scheme.

can be obtained by paying attention to that the charges accumulated in the upper electrodes of the capacitors are preserved in the transition from Figure 5.22(a) to Figure 5.22(b). If $V_{x1} < 0$, which means that $-V_{in} + \frac{1}{2}V_{ref} < 0$, then MSB $= 1$, and connecting the capacity as shown in Figure 5.22(c) yields the second MSB. If $-V_{in} + \frac{1}{2}V_{ref} > 0$, then MSB $= 0$, and connecting the capacity as shown in Figure 5.22(d) yields the second MSB. By repeating these steps, the circuit can perform the binary search algorithm shown in Figure 5.17.

5.4.3 Energy Consumption

5.4.3.1 Conventional capacitor array

Consider quantitatively the energy consumption of capacitor-based DACs used in SAR A/D converters [68–70]. The energy consumption instead of the power consumption is appropriate to focus each sampling event in A/D conversion. The power consumption can be estimated by multiplying the energy by the sampling frequency.

Figure 5.23 shows the model to estimate the energy consumption, which can be calculated as the work done by the voltage source V_{ref} when the capacitors are charging or discharging. The energy required for the MSB decision after sampling, $E_{samp \to 1}$, is obtained as[8]

$$
E_{samp \to 1} = \int_0^1 I_{ref}(t) V_{ref} dt = V_{ref} \int_0^1 \frac{dQ_{2C}}{dt} dt = V_{ref} \int_{Q_{2C}(0)}^{Q_{2C}(1)} dQ_{2C}
$$

$$
= V_{ref} [2C(V_{ref} - V_{x1}) - 2C(V_{in} - 0)] = CV_{ref}^2. \tag{5.9}
$$

Next, when MSB $= 0$, the energy $E_{1 \to 20}$ necessary for the second MSB decision can be obtained as

$$
E_{1 \to 20} = V_{ref} \int_{Q_C(0)}^{Q_C(1)} dQ_{2C} = V_{ref} [C(V_{ref} - V_{x20}) - C(0 - V_{x1})]
$$

$$
= \frac{5}{4} CV_{ref}^2. \tag{5.10}
$$

---

[8]The electrostatic energy $(1/2)CV_{ref}^2$ is half of the value obtained by Equation (5.9). The other half is the Joule heat consumed by the resistor. When charging or discharging a capacitor with a constant voltage source, it is necessary to take into consideration the resistor for absorbing the difference between the power supply voltage and the capacitor voltage. The resistor here is the on-state MOSFET switch.

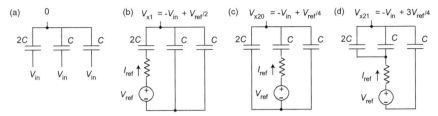

Figure 5.23   Circuit models to estimate the energy consumed by the reference voltage source. (a)–(d) illustrate modes of sampling, MSB decision, second MSB decisions when $MSB = 0$, and second MSB decisions when $MSB = 1$, respectively.

On the other hand, when $MSB = 1$, the energy $E_{1\to21}$ necessary for the second MSB decision is

$$E_{1\to21} = V_{ref}\left[2C(V_{ref} - V_{x21}) - 2C(V_{ref} - V_{x1})\right]$$
$$+ V_{ref}\left[C(V_{ref} - V_{x21}) - C(V_{in} - 0)\right] = \frac{1}{4}CV_{ref}^2. \quad (5.11)$$

Figure 5.24 shows the energy consumption required for determining each bit in a 4-bit D/A converter. Energy values are normalized with CV_{ref}^2. The energy required for A/D conversion with a 4-bit resolution is obtained by adding up the energy consumed in each bit decision, which depends on the output code. The result is plotted by $\times$'s in Figure 5.25. The energy consumption is the largest when the output code is 0000. This is attributable to a large number of switching operations for discharging capacitors in the case of 0000. For example, when V_{in} is smaller than $V_{ref}/2$, the threshold voltage for the bit decision needs to be changed from $V_{ref}/2$ to $V_{ref}/4$. Reducing the decision voltage for comparison is referred to as a downward transition. At this time, it is necessary to charge $C_3(= 4C_0)$ after discharging $C_4(= 8C_0)$ in Figure 5.24, leading to consuming energy. On the other hand, when $V_{ref}/2$ is smaller than V_{in}, the decision voltage is changed from $V_{ref}/2$ to $3V_{ref}/4$. This is called an upward transition. $C_3(= 4C_0)$ is charged while $C_4(= 8C_0)$ is kept being charged. This means that less energy is needed than the downward transition. If a part of charges in capacitor $C_4(= 8C_0)$ can be used to charge $C_3(= 4C_0)$ in the downward transition, the energy consumption in the downward transition becomes small. The method described below realizes low-power conversion based on this idea.

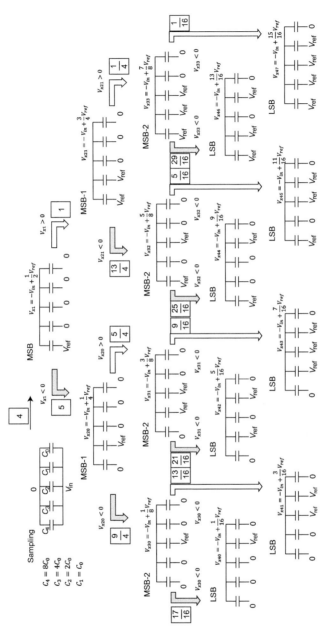

Figure 5.24 Operation and energy consumption in a 4-bit D/A converter using a binary-weighted capacitor array.

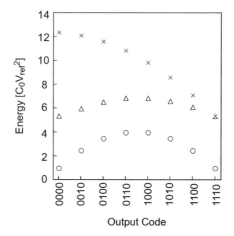

Figure 5.25   Comparison of energy consumption of capacitor-based DACs used in SAR ADCs. ×'s, △'s, and ○'s represent the results for a conventional-type binary-weighted capacitor scheme, split capacitor scheme, junction splitting scheme, respectively.

5.4.3.2 Split-capacitor array

A split-capacitor D/A converter was proposed to reduce the energy consumption associated with the downward transition [69]. All capacitors other than the one corresponding to the LSB are split into halves. The switching operation of a 2-bit SAR A/D converter using this scheme is shown in Figure 5.26. C_{21} and C_{22} correspond to $2C$ in Figure 5.23. In the conventional MSB decision, C_{21} and C_{22} are first connected to V_{ref}, but in the split-capacitor scheme, one of them, C_{21}, is connected to V_{ref}. Also, C_1 instead of C_{22} is connected to V_{ref}. Since the sum of these capacitors is equal to $2C$, the charge distribution is effectively the same as in the conventional scheme, and V_{in} can be compared with $V_{\text{ref}}/2$. In determining the second bit, C_1 remains connected to V_{ref} regardless of whether the MSB is 0 or 1. Therefore, in contrast to the conventional scheme, where all the charges of $2C$ are discarded, half of them are reused to save energy.

Switching operation and energy consumption of a 4-bit SAR A/D converter using a split-capacitor DAC are shown in Figure 5.27. The energy consumption is calculated in the same manner as in the case of the conventional DAC explained above. The result is indicated by △'s in Figure 5.25. The energy can be efficiently used in the left half region corresponding to $\text{MSB} = 0$, where the downward transitions take place more frequently than in the right half region. Since the total capacitance value is

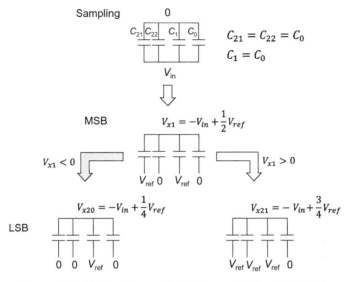

Figure 5.26 2-bit split-capacitor DAC operation in a SAR ADC.

the same as that in the conventional type, the occupied area is almost the same. The drawback is that the number of switches almost doubles, and the logic circuit to control them and the wiring become complicated.

5.4.3.3 Junction-splitting array

In the conventional D/A converter and the split-capacitor D/A converter described above, the energy required for determining the MSB is more than that required for determining any other subsequent bits. This is because the capacitor used for the MSB decision is the largest. If smaller capacitors can be used for the MSB decision, energy consumption can be reduced. Based on this idea, a junction-splitting (JS) D/A converter was proposed [71], the switching operation of which is shown for a 2-bit case in Figure 5.28. The capacitor configuration is basically the same as the conventional one, but C_2 is disconnected from C_1 and C_0 in the MSB decision. When the MSB is decided, the switch is closed for the LSB decision. Once the capacitor has been charged, it will not be discharged during the subsequent decision operation.

Each switch operation and energy consumption in a 4-bit SAR A/D converter based on the junction-splitting scheme are shown in Figure 5.29. Also, the energy necessary for the 4-bit resolution is indicated by ○'s

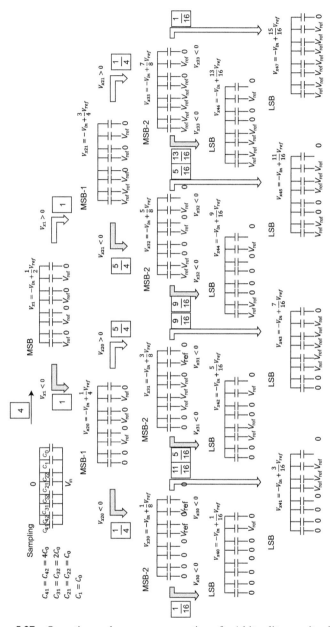

Figure 5.27   Operation and energy consumption of a 4-bit split-capacitor DAC.

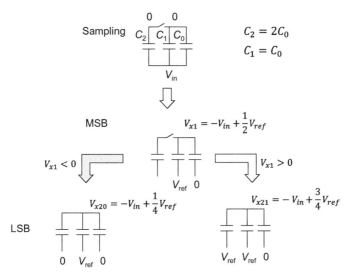

Figure 5.28 2-bit junction-splitting DAC operation in a SAR ADC.

in Figure 5.25. It should be noted that MSBs are compared with small capacitors. Therefore, the MSB decision is more sensitive to variations in capacitance values than the conventional type and the split capacitor scheme, which decide MSBs using large capacitors.

Typical examples for reducing the energy consumption associated with capacitor-array charging and discharging have been mentioned, which is critical in pursuing low-power consumption in SAR A/D converters. Even in these days, the sophistication of the switching sequence is progressing for further reduction of the energy consumption. As a result, the energy consumption of other circuit blocks in the SAR A/D converters, such as a comparator, successive approximation register (SAR), and switch control logic, becomes comparable with that of the capacitor array. In the current SAR A/D converter, roughly speaking, the analog circuit, including the D/A converter and a comparator, consumes almost half of the total power, and the digital circuitry does the other half. Also, note that the power consumption increases in proportion to sampling frequency.

A technique called top plate sampling [72] is also known for achieving lower-energy operation. By directly applying V_{in} to the comparator input terminals to determine the MSB, it is possible to halve the required total capacitance value, thus enabling low-energy operation. However, for the

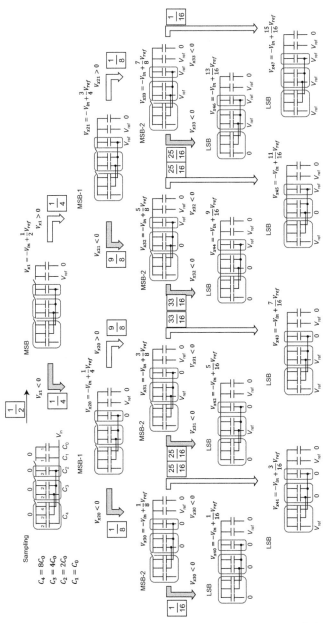

Figure 5.29 Operation and energy consumption of a 4-bit junction-splitting DAC.

sampling, the input voltage is applied to the upper electrodes of the capacitors, and for the following bit decision, V_{ref} is applied to the bottom electrodes. Since the parasitic capacitance associated with these electrodes are different from each other, the threshold of the bit decision might also change, and sufficient attention is required in designing circuits and drawing the layout. Another low-energy configuration called monotonic switching was also proposed [73]. However, since the common-mode input voltage to the comparator can change in a wide range from ground to V_{DD}, it is necessary to ensure linearity over a wide range of input.

Up to now, it is assumed that a single bit is determined by each cycle in successive approximation. If a 2-bit flash A/D converter is used instead of one comparator described above, 2 bits can be determined in each cycle, which reduces the required number of cycles by half. This is known as a 2-bit/cycle scheme [74, 75]. Combining this with the top-plate sampling makes it possible to reduce energy consumption further. However, unlike 1 bit/cycle, it is necessary to generate multiple threshold voltages for each comparison, making the circuit more complicated.

5.4.4 Decision with Redundancy

In the explanation above, no fluctuation was assumed in capacitor values in D/A converters, and the values are accurate as designed. However, in reality, variations in the capacitance value, or capacitance mismatches, inevitably occur. An example is shown in Figure 5.30. Here, it is assumed that the MSB and the second MSB are 0 and 1, respectively, and that the capacitor for the LSB decision is $C + \Delta C$. The D/A-converter output for the LSB decision is then expressed as

$$V_{\text{x}} = \frac{1}{1 + \alpha/8} \left\{ -V_{\text{in}} + \frac{3}{8} \left(1 + \frac{\alpha}{3} \right) V_{\text{ref}} \right\}. \tag{5.12}$$

Here, α is a relative error equal to $\frac{\Delta C}{C}$. The mismatch results in an incorrect output of 010 instead of 011, which is the output obtained when there is no mismatch ($\alpha = 0$). Similarly, an incorrect decision can occur in the MSB if such a deviation occurs for $4C$. Then, bit decisions following the MSB would make no sense. In other words, if the wrong decision takes place, it can never be corrected. This is due to the binary representation without redundancy.

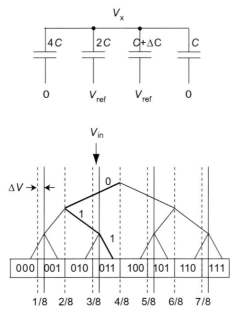

Figure 5.30 Bit decision error due to capacitance mismatch. ΔV is a variation in the threshold voltage due to mismatch.

In contrast, by introducing redundancy, such a wrong decision can be recovered as shown below. Let us consider a number representation using a radix R smaller than 2. $(D_0 D_1 D_2 \cdots D_N)_R$ is represented by a decimal number as

$$(D_0 D_1 D_2 \cdots D_N)_R = \sum_{i=0}^{N-1} D_i R^{-1-i}. \tag{5.13}$$

The representation has redundancy: For example, if $R = 1.75$, both $(100010)_{1.75}$ and $(011111)_{1.75}$ are approximately equal to 0.72 in the decimal representation. This means that the same numerical value in the decimal representation cannot be uniquely determined in a number system with a radix smaller than 2. When this is applied to the SAR A/D converter, even if the MSB decision is incorrect, the subsequent bit decision can correct the error.

Figure 5.31 schematically illustrates such a redundant decision. If the MSB is correctly determined for the input value V_{in}, the output is 1000. On the other hand, if the MSB decision is erroneous to become 0, continuing the

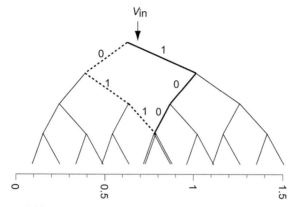

Figure 5.31 Bit decision process in a SAR ADC with redundancy.

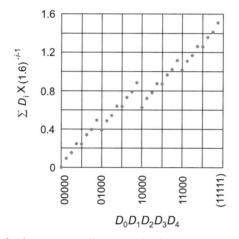

Figure 5.32 Decimal value corresponding to a redundant representation with a radix of 1.6.

subsequent determination will result in 0110, which will be approximately the same as 1000 if these are represented in decimal notation. Continuing to determine the lower bits further reduces the error, which means that errors in the MSB decision can be corrected [76]. Figure 5.32 shows the relation between the five-digit codes in a radix of 1.6 and the corresponding decimal numbers. There is a gap, for example, at 10000, which represents the overlapped part of Figure 5.31, or the redundancy. Note that it is necessary to increase the number of cycles to obtain the necessary bit resolution.

The redundant decision has another advantage. It is known that the conversion speed can be increased even if the number of cycles increases [77].

In the SAR conversion, the bit decision should be made after completing the settling in the DAC capacitor array, which can sometimes last longer. However, the correction based on the redundancy makes it possible to decide the bit without waiting for the settling to complete, because even if an erroneous decision is made as a result of insufficient settling, the error can be corrected by subsequent decisions. Therefore, the sampling frequency can be made higher than that in the conversion without redundancy.

Note that if $R < 2$ is adopted as the radix, the capacitance ratio in the capacitor array becomes non-integer. Recall that to use capacitors weighted by integer ratio is highly desirable to reduce mismatches as mentioned in Section 4.4, and non-integer weighting should be avoided. There exist several proposals to solve the problem, and one of them is known as a generalized non-binary SAR A/D converter [78]. It is a method of giving redundancy while maintaining the integer ratio. For example, to obtain an 8-bit resolution, conventionally, the binary-weighting factors are 128, 64, 32, 16, 8, 4, 2, and 1. In the generalized non-binary scheme, by selecting the weighting factors as, for example, 128, 46, 26, 20, 14, 8, 6, 4, 2, 2, and 1. The redundancy is taken into consideration by increasing the number of weighting factors, and a wrong decision can be corrected. Note that the total capacitance value does not change, but the required number of cycles increases.

If an N-bit resolution is obtained by M cycles, the output D_{out} is expressed as [78]

$$D_{\mathrm{out}} = 2^{N-1} + \sum_{i=2}^{M} s(i-1)p(i) + \frac{1}{2}\left(s(M) - 1\right), \qquad (5.14)$$

where $p(k)$ is the weighting factor that is added to or subtracted from the reference voltage in the previous cycle, and $s(i)$ is defined as

$$s(i) = \begin{cases} -1 & (d(i) = 1) \\ 1 & (d(i) = 0) \end{cases}. \qquad (5.15)$$

Here, $d(k)$ is the result of the k-th decision.

An example of redundant decisions in the generalized non-binary algorithm is shown in Figure 5.33. Here, a 5-bit resolution ($N = 5$) and a cycle number of 6 ($M = 6$) are assumed. $d(1)$ is the MSB, $d(6)$ is the LSB. $p(i)$ is chosen as follows: $p(1) = 16$, $p(2) = 7$, $p(3) = 4$, $p(4) = 2$, $p(5) = p(6) = 1$. If the input is $V_{\mathrm{in}1}$, the correct output is 100010. When an incorrect value of 0 is decided as the MSB, 011110 is obtained as the output. With Equation (5.14), in either case, a decimal number of 16 is obtained,

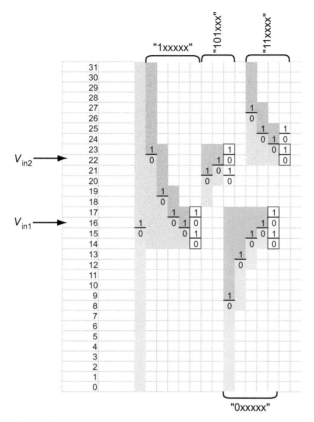

Figure 5.33   Illustration of a generalized non-binary search.

which agrees with the input. Also, when the input is V_{in2}, the correct output is 101110, but if an incorrect decision was made in the second cycle, it is 110000. The output calculated using Equation (5.14) is 22 in either case.

As described above, since the SAR A/D converter does not require a power-hungry amplifier, it is suitable for low-power operation. Recently, a low-power small-area SAR ADC was reported for CMOS image sensors [79]. Also, a new class of SAR ADCs using a rotating-and-averaging method [80] and a reconfigurable capacitor DAC [81] were proposed to enhance the linearity instead of introducing redundancy. Further reduction of the power consumption is intensively pursuit in these days [82, 83]. As explained in Chapter 7, future improvements in the bit resolution are expected by using digital correction, noise shaping, and pipelining. For interested readers, please refer to the articles [84–87].

5.5 Two-step/Subranging/Algorithmic A/D Converters

In flash A/D converters and folding-and-interpolation A/D converters, all the bit decisions are processed in parallel, or a single step, using a large number of comparators. In contrast, multi-step A/D converters perform the conversion for the upper bits first and then use the quantization error to determine the next upper bits. This procedure continues until the lowest bit is obtained. A typical example of a two-step A/D converter is shown in Figure 5.34. Two sub-A/D converters, ADC_1 and ADC_2, are usually flash A/D converters. Other types of A/D converters can also be used as sub-A/D converters. Usually $N_1 + N_2 = N$, but if redundant decisions are introduced, $N_1 + N_2 > N$. The gain A is typically 2^{N_1} so that the input full-scale of ADC_2 is the same as that of ADC_1. Amplifying the quantization error is schematically shown in Figure 5.35. As shown in the figure, 1 LSB in the upper 2-bit decision is amplified to the full scale to determine the lower 2-bits.

A similar type of an A/D converter, called a subranging A/D converter, is shown in Figure 5.36. Here, the result in the first stage (ADC_1) determines the full scale in the next stage (ADC_2). Although sometimes the structure shown in Figure 5.34 is called a subranging type, it is desirable to distinguish them clearly.

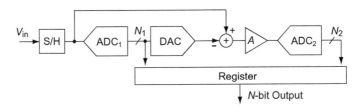

Figure 5.34   Block diagram of a two-step ADC.

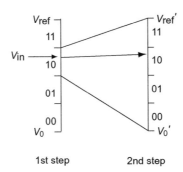

Figure 5.35   Amplifying quantization error in a 2-step ADC.

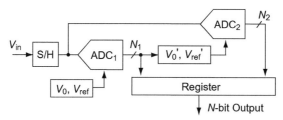

Figure 5.36 Block diagram of a subranging ADC.

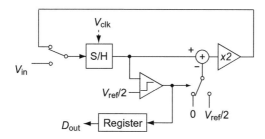

Figure 5.37 Block diagram of an algorithmic ADC.

Instead of spatially arranging the conversion units like a two-step type shown in Figure 5.34, it is also possible to adopt a feedback loop to realize multi-stage processing in a time-division manner as shown in Figure 5.37. The sub-A/D converter used in the loop is replaced with a 1-bit A/D converter or a comparator. This is called an algorithmic or cyclic A/D converter. It looks similar to the SAR A/D converter in that the output can be obtained bit by bit from the MSB to the LSB, but the difference is that the signal is amplified by using the amplifier, while the threshold is constant, which is $V_{\mathrm{ref}}/2$ in this example. There is an advantage that the occupied area can be small because the same circuit unit is repeatedly used instead of arranging multi-stage sub-A/D converters on a Si substrate.

The advantage of the A/D converters described in this section is that higher resolution can be achieved with a small circuit scale without using folding and interpolating circuits. However, it should be noted that the time required for conversion increases and power efficiency is not as good when opamps are used.

5.6 Pipelined A/D Converters

A pipelined A/D converter is a variation of multi-step A/D converters which is equipped with a S/H circuit in each stage to enable pipelined

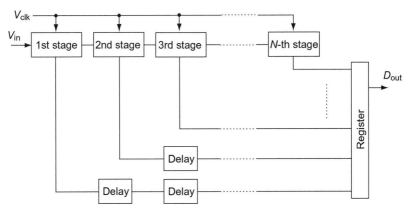

Figure 5.38   Block diagram of a pipelined A/D converter.

operation [88, 89][9]. It consists of N stages of sub-A/D converters. The quantization error generated in the preceding stage is amplified and applied to the next stage. The last stage uses a flash A/D converter. The output of each stage is delayed and added to obtain the final output of D_{out}. Thus, the pipelined A/D converter can be thought of as an algorithmic A/D converter spatially extended on a silicon chip as shown in Figure 5.38.

The block diagram of each stage of the pipelined A/D converter is shown in Figure 5.39. In this example, it is assumed that the sub-A/D converter has a 2-bit resolution. V_3 is the difference between the analog input and the corresponding quantized output, which is a quantization error and called a "residue." The amplifier is called a residue amplifier. The gain is determined so that 1 LSB of the present stage becomes the full scale of the next stage. For example, when the resolution of the sub-A/D converter is 2 bits, the residue-amplifier gain needs to be precisely 4. Although the output is 2 bits, the conversion precision corresponding to the total bit resolution after that stage is required. This means that conversion at the first stage should be done with the highest accuracy. In Figure 5.39, the dotted frame, including the subtractor, multiplier, and D/A converter, is called a multiplying DAC (MDAC).

The time it takes from sampling an analog signal to generating the final digital code is called latency. Latency increases as the number of stages increases. However, once the first digital code is obtained, the following outputs are obtained at the rate of V_{clk}. V_{clk} is determined by the conversion

---

[9]This is the same idea to speed up used in digital processors.

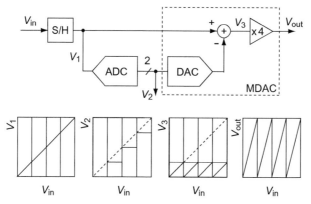

Figure 5.39 Block diagram of a sub-ADC for each stage of a pipelined A/D converter.

time of the sub-A/D converter so that the high-speed operation in this sense can be expected for the entire pipeline. Thus the throughput of pipelined A/D converters approaches that of flash converters. In application fields where real-time response is the priority, a long latency cannot be allowable. On the other hand, for example, for a receiver whose data rate is predetermined, if the throughput is high enough, the latency is not a big problem. Therefore, the pipelined A/D converter is widely used in such communication applications.

Consider the relation between the bit resolution of the sub-A/D converter and the number of stages required to obtain a specific bit resolution. If the bit resolution in each stage increases, the number of stages can be reduced. Conversely, the required number of stages increases if the bit resolution is reduced to simplify the sub-A/D converter. It depends on constraints on the circuit scale, power consumption, latency, and throughput whether to increase or decrease the bit resolution of the sub-A/D converter. If the bit resolution increases, a high gain is required for the residue amplifier. As mentioned in Section 3.2.1, increasing the gain slows the amplifier operation, and the power consumption increases. According to a systematic analysis [90], it is possible to reduce the conversion speed, power consumption and occupied area by reducing the bit resolution of each stage[10].

A circuit diagram of a 1-bit sub-A/D converter is shown in Figure 5.40. ϕ_1 represents the sampling mode while ϕ_2 is for the amplification mode. A circuit diagram corresponding to each mode is shown in Figure 5.41. In the

---

[10]Note that this conclusion depends on the technology assumed. In recent years, contrary to this, a two-stage configuration using a SAR A/D converter as a sub-A/D converter has attracted attention as will be described in Section 7.3.

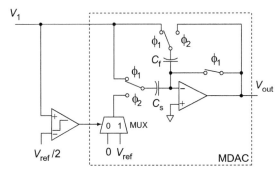

Figure 5.40   Circuit diagram of a 1-bit sub-ADC.

(a)

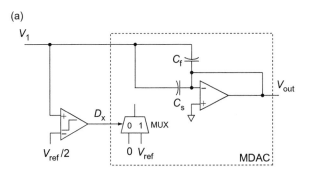

(b)

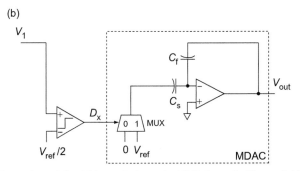

Figure 5.41   Operation of the 1-bit resolution sub-ADC shown in Figure 5.40: (a) sampling mode and (b) amplification mode.

sampling mode (Figure 5.41(a)), the input voltage V_1 is stored as the charge in two capacitors C_s and C_f. Their upper electrodes are virtually grounded. In the amplification mode shown in Figure 5.41(b), C_f is in the feedback path and the charges are redistributed. The potential of the bottom electrode of C_s is either 0 or V_{ref} depending on the output of the comparator D_x.

The output V_{out} can be written as

$$V_{\text{out}} = \frac{C_s + C_f}{C_f} V_1 - \frac{C_s}{C_f} V_{\text{ref}} D_x. \tag{5.16}$$

If $C_s = C_f$,

$$V_{\text{out}} = \begin{cases} 2V_1 & (D_x = 0) \\ 2V_1 - V_{\text{ref}} & (D_x = 1) \end{cases} \tag{5.17}$$

is obtained. If there is a capacitance mismatch between C_s and C_f, the coefficient of V_1 deviates from 2. This is called a gain error. If there is an offset in the comparator, the input value corresponding to the boundary where the output changes from 0 to 1 deviates from $V_{\text{ref}}/2$. This is an offset error.

According to Equation (5.17), the transfer characteristics of the 1-bit sub-A/D converter are shown in Figure 5.42. Solid lines in Figures 5.42(a) and (b) show characteristics without and with the offset error, respectively. In these figures, the dotted lines show the characteristics obtained by exchanging the input and output axes. This enables to trace the output from each stage. If there is no offset error, the output is obtained as $0110 \cdots$ for the input V_{in1} as shown in Figure 5.42(a). If there is an offset error (Figure 5.42(b)), an incorrect code of $1000 \cdots$ is generated for the same input V_{in1}.

Even if an incorrect decision due to the offset error occurs, it can be corrected by using a digital correction technique [91]. One of the most common approaches is a 1.5-bits/stage technique, the circuit diagram of which is shown in Figure 5.43. The input range is divided into three parts

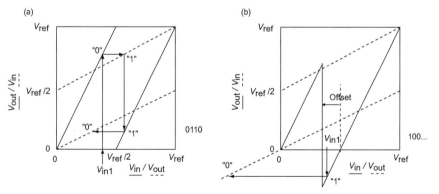

Figure 5.42 Bit decisions in a 1-bit/stage pipelined A/D converter. (a) Normal decision and (b) error decision due to threshold variation.

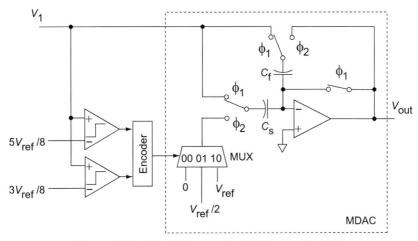

Figure 5.43   Circuit diagram of a 1.5bit/stage sub-ADC.

by using two comparators, each of which is denoted as 00, 01, or 10. In this case, the output V_{out} can be written as

$$V_{\text{out}} = \begin{cases} 2V_1 & (V_1 < 3V_{\text{ref}}/8) \\ 2V_1 - V_{\text{ref}}/2 & (3V_{\text{ref}}/8 \le V_1 < 5V_{\text{ref}}/8) \\ 2V_1 - V_{\text{ref}} & (5V_{\text{ref}}/8 \le V_1) \end{cases} \qquad (5.18)$$

Here $C_{\text{s}} = C_{\text{f}}$ is assumed. The input/output characteristics are shown in Figure 5.44(a). For an input of $V_{\text{in}1}$, the decision result is 01, followed by 00, 01, and 10. As written on the right of Figure 5.44(a), 01100 is obtained as an output by shifting the digits and adding them together. On the other hand, if there is an offset as shown in Figure 5.44(b), 00, 10, 01, and 10 are generated. In the same manner as mentioned above, 01100 is obtained as the output. Therefore, the two outputs are the same. If the offset is within $\pm V_{\text{ref}}/8$, it is possible to correct the decision error due to the offset[11].

Figure 5.45 illustrates the redundant decision in the 1.5-bit/stage pipelined A/D converter. For example, when the input is $V_{\text{in}1}$, the first decision is 10, followed by 00, 01, and 10. Shifting the digits and adding them, as shown in Figure 5.44, results in 10100 in binary or 20 in decimal. On the other hand, when the first decision is 01 by error instead of 10, this is followed by 10, 01, and 10, also resulting in 20 in decimal. Thus the decision error is corrected.

---

[11]Notice that this method cannot correct the gain error, and another digital correction technique described in Section 7.4 is necessary.

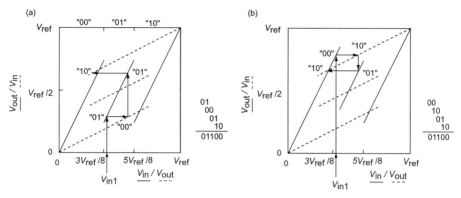

Figure 5.44 Digital correction in a 1.5bit/stage pipelined A/D converter. (a) Normal operation and (b) example of correction of a wrong decision due to threshold deviation.

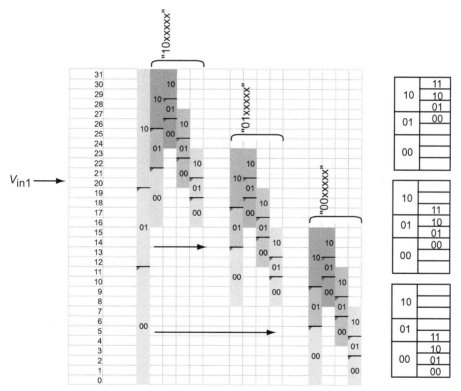

Figure 5.45 Illustration of bit decision in a 1.5-bit/stage pipelined ADC.

The output of 1.5 bits/step corresponds to a signed-digit binary number representation [92] in which values of -1, 0, 1 are allowed for each digit. Now assume that t_i can take values of $-1, 0, 1$, and that y is represented as

$$y = \sum_{i=1}^{4} t_i 2^{-i}. \tag{5.19}$$

Then, when y is decided, the corresponding t_i is not uniquely determined. In fact, in the example of Figure 5.44(a), by replacing 00, 01, 10 with -1, 0, 1, respectively, $t_1 = 0$, $t_2 = -1$, $t_3 = 0$, and $t_4 = 1$ results in $y = -3/16$. This value is also obtained by assuming $t_1 = -1$, $t_2 = 1$, $t_3 = 0$, and $t_4 = 1$, which correspond to the output in Figure 5.44(b). Therefore, it is understood that by adopting the redundancy as was described for the SAR A/D converter, decision errors can be digitally corrected.

The pipelined architecture has been widely used in A/D converters capable of obtaining a high resolution with a medium- to high-speed range [93–98]. High-speed performance has been further improved by using them in a time-interleaved manner [99, 100]. For example, by operating eight pipelined A/D converters with a 12-bit resolution in parallel, a sampling rate of 10 GS/s was achieved [101]. On the other hand, in the low-resolution range, the performance is competing with SAR A/D converters. Especially, from the viewpoint of low power consumption, the pipelined type is inferior. Furthermore, in the high-resolution region, it competes with oversampling $\Delta\Sigma$-type A/D converters to be described in Chapter 7.

The use of amplifiers has become a bottleneck in reducing power consumption. Besides, the advance of device miniaturization makes it challenging to design high-performance amplifiers [102]. To solve this problem, research on a new class of amplifiers which may replace conventional operational amplifiers, such as a dynamic amplifier [103] and a ring amplifier [104] is progressing. These amplifiers are described in Section 7.2.

5.7 Integral/Time-domain A/D Converters

In A/D converters described above, the voltage represents analog signals, and the bit resolution depends on how small the full-scale voltage can be divided. If the full-scale voltage is several volts, it is necessary to divide the signal with a several mV precision to obtain a 10-bit resolution and a few μV precision for 20 bits. Now, the digital LSI technology has been progressing towards lower-voltage operation than before. If data converters also follow this trend, it is inevitable that the available voltage amplitude decreases and it becomes increasingly difficult to achieve a high bit resolution. Also, it is not easy to design high-performance amplifiers with advanced CMOS technology.

Under these circumstances, in recent years, processing the signal in the time domain began to draw attention. Here, the analog input voltage is converted into the time-domain signal, and then the signal is digitally evaluated. This approach is attractive not only because the low-voltage operation is possible, but also because the high-speed operation based on scaled-down CMOS technology is expected to enhance the resolution in the time domain. In this section, A/D converters based on signal processing in the time domain are described. First, an integration type that has long been known as A/D converters with high resolution but not high speed is explained. Next, A/D converters using the time-to-digital converter (TDC) is described.

5.7.1 Integration Type

The block diagram and the timing chart of an integration-type A/D converter are shown in Figure 5.46. V_{start} activates the ramp-wave generator, and at the same time, it starts counting pulses from the oscillator. The ramp-wave generator is an integration circuit using an opamp. The slope of the increasing output of V_x is proportional to the input voltage V_{in}. Since the input voltage is integrated, such an A/D converter is called an integration type. If V_x becomes equal to V_{ref} at t_1, the comparator output becomes low, and the counter stops counting. V_x can be written as

$$\frac{dV_x}{dt} = \frac{1}{C}\frac{dQ_x}{dt} = \frac{1}{C}\frac{V_{\text{in}}}{R}. \tag{5.20}$$

Since its initial value is 0 V,

$$V_x = \frac{V_{\text{in}}}{CR}t \tag{5.21}$$

is obtained, and

$$V_{\text{in}} = \frac{CR}{t_1}V_{\text{ref}} \tag{5.22}$$

is satisfied at t_1. If the oscillator frequency is known, t_1 can be obtained as a digital value by the counted number of pulses. Since V_{ref}, R and C are also known, digital output equivalent to V_{in} is obtained by Equation (5.22).

An alternative ramp-wave generator is shown in Figure 5.47 [105]. This consists of a cascode current mirror to charge the capacitor with a constant

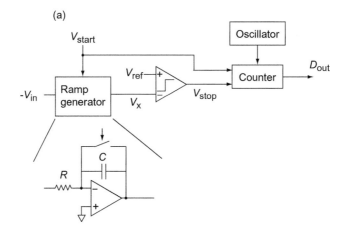

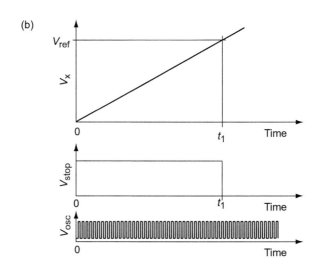

Figure 5.46 (a) Block diagram and (b) voltage waveforms in a single-slope integration-type ADC.

current. Generating a highly linear ramp wave is critical because it determines the conversion linearity.

Equation (5.22) indicates that the A/D conversion needs accurate values of the resistance and capacitance of the ramp-wave generator. In the standard CMOS process, however, absolute values of the resistance and capacitance

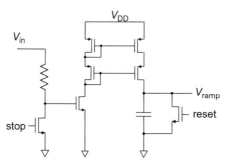

Figure 5.47   Ramp wave generator.

fluctuate about ±10%. Thus, it is difficult to achieve high accuracy with this method. Another integration-type A/D converter that solves this problem is shown in Figure 5.48. This is called a dual-slope integration-type A/D converter, or simply a dual-slope A/D converter. To distinguish from this, the one illustrated in Figure 5.46 is called a single-slope integration-type A/D converter.

Operation of the dual-slope A/D converter is as follows. At $t = 0$, the ramp-wave generator (Ramp: Up) starts with an initial value of 0 V by the signal V_{start}. The slope of the ramp wave is proportional to the input V_{in}. At $t = T_1$, the output voltage of the generator increases up to V_{up}. At the same time, the second ramp-wave generator (Ramp: Down) starts with an initial value of V_{up}, and the counter starts pulse counting. The negative slope is a constant determined by V_{ref}. The comparator monitors the output of the ramp generator and stops the counting when $V_{\text{down}} = 0$. Since the peak voltage of the ramp wave is proportional to the input voltage, and the following negative slope is constant, the time from T_1 to the time when the comparator output (t_1 to t_3) becomes 0 is proportional to the input. Therefore, the number of counted pulses represents the digital output.

A dual-slope waveform shown in Figure 5.48(b) can be obtained by using the ramp-wave generator shown in Figure 5.46(a) by applying $-V_{\text{in}}$ at $t = 0$, and then V_{ref} at $t = T_1$. The ramp wave voltage before T_1 can be expressed as

$$\frac{V_{\text{in}}}{R} = I = \frac{dQ}{dt} = C\frac{dV}{dt}. \tag{5.23}$$

If the zero crossing occurs at t_{x}, then

$$\frac{V_{\text{in}}}{RC}T_1 = \frac{V_{\text{ref}}}{RC}t_{\text{x}} \tag{5.24}$$

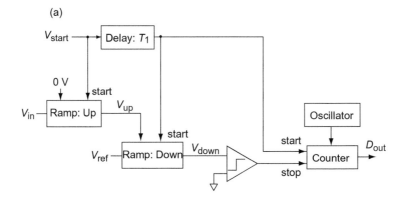

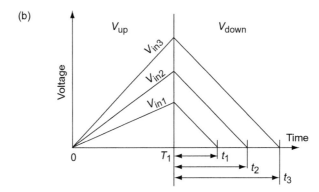

Figure 5.48 (a) Block diagram and (b) voltage waveforms in a dual-slope integration-type ADC ($V_{in1} < V_{in2} < V_{in3}$).

holds. Therefore,

$$t_x = \frac{V_{in}}{V_{ref}} T_1 \tag{5.25}$$

is obtained. Now, t_x is determined only by V_{in} and V_{ref}, and is independent of the resistance and capacitance values, which means that these absolute values do not affect the conversion accuracy. This is because charging and discharging are performed by using the same capacitor and resistor.

Compared with other A/D converters, the configuration of the integration-type A/D converter is compact, and the occupied area can be small. An example which takes advantage of the small area is a column A/D converter in an image sensor, where a large number of A/D converters are used for each column of the pixel matrix. Also, in the integration-type A/D converter, high bit resolution can be realized by increasing the integration time. However, to do so, it is necessary to count many pulses. For example, to obtain a 20-bit resolution, $2^{20} (\approx 10^6)$ pulses are required. If the oscillator frequency is 1 GHz, it would take approximately 10^{-3} seconds. This means that the upper limit on the sampling frequency is only 1 kHz. Therefore, this type of A/D converters is suitable for high-resolution but low sampling-frequency applications. Hybrid configurations combined with other schemes are worth being considered to improve the operation speed.

A unique feature of the integration-type A/D converter is that it can reject the noise if its frequency is equal to integer multiples of the integration time. For example, this is quite effective to suppress the noise caused by the commercial power supply.

5.7.2 Time-to-digital Type

In the integration type described above, an analog circuit was used to express the voltage in the time domain. This subsection explains another A/D converter, which use a signal delay line consisting of a digital circuit to convert the voltage to the time difference. The time difference is then converted into a digital output. The circuit used for converting the time difference into a digital value is called a time-to-digital (T/D) converter or TDC.

A digital signal delay line uses an inverter (a NOT gate) as shown in Figure 5.49. When a start pulse is applied to the left end, the signal is delayed by τ after passing through the delay elements consisting of two inverters. The signal delay time of the inverter depends on the power supply voltage. If the supply voltage is high, the delay time becomes short, and vice versa. Therefore, if the analog input V_{in} is applied as the supply voltage, V_{in} can be estimated by measuring τ.

If both ends of the delay line shown in Figure 5.49 are connected to form a ring oscillator, it is possible to estimate a time longer than τ by counting the number of oscillations. Since the voltage can control the oscillation frequency, this is called a voltage-controlled oscillator (VCO), and an A/D converter using the VCO as T/D converter is sometimes called a VCO-based A/D converter. It attracts increasing attention because almost all circuits are digital, and this is compatible with scaled-down CMOS technology. Such a simple ring oscillator, however, has the disadvantage that the allowable input range is narrow. As a modified VCO that can handle a wider range of V_{in} is

(a)

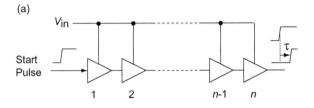

(b)

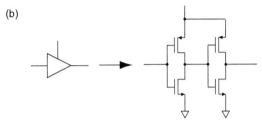

Figure 5.49   (a) Delay line and (b) delay element consisting of inverters.

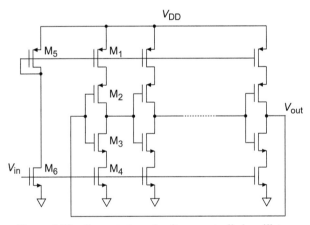

Figure 5.50   Current-starved voltage-controlled oscillator.

shown in Figure 5.50, which is called a current-starved VCO [28]. M_2 and M_3 constitute an inverter. M_1 and M_4 determine the current flowing through the inverter. V_{in} is applied to the current mirror consisting of M_1, M_5, and M_6. When V_{in} increases, the current flowing through the inverters increases, and the oscillation frequency increases, so that the digital equivalent to V_{in} can be obtained by measuring the oscillation frequency.

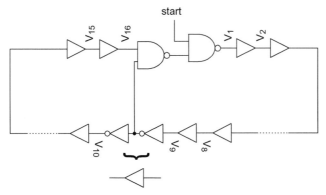

Figure 5.51   Ring oscillator used in a time domain A/D converter.

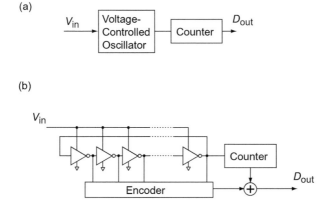

Figure 5.52   (a) Time-domain A/D converter and (b) circuit implementation.

A ring oscillator used in time-domain A/D converters is shown in Figure 5.51 [106]. In this circuit, it operates as a typical ring oscillator when the start signal is HIGH. When the start signal goes LOW, V_1 is fixed at HIGH, and it reaches V_{16}. As a result, all nodes become HIGH, oscillation stops, and the oscillator enters the reset state. Therefore, counting always starts from the same initial state[12].

A block diagram of a time-domain A/D converter using a VCO is shown in Figure 5.52 [108]. The counter shows how many times the signal has rotated around the ring. Also, if the node where 0 changes to 1 can be detected

---

[12]If you do not reset, the quantization error is taken over to the next sampling, and the first-order noise shaping is obtained [107].

by the encoder, resolution smaller than one cycle time can be obtained. In other words, at the same time as the number of oscillations, it can be determined to which part of the delay line the signal has advanced. Digital outputs are then generated based on them. Note that there is nonlinearity between the input voltage and the oscillation frequency in the VCO. In order to avoid this influence, limiting the input voltage range only to the portion with excellent linearity is one possible solution. It is also possible to perform digital correction using a memory storing the non-linear characteristics.

5.8 Time-interleaved A/D Converters

With advances in scaled-down CMOS technology, it becomes possible to fabricate many A/D converters on a Si substrate and to operate them in parallel like a multi-core processor. It is called a time-interleaved A/D converter [109]. Parallel operation enables fast A/D conversion even though the individual operation is slow. For example, a time-interleaved SAR A/D converter operates with a conversion rate comparable to pipelined and flash A/D converters with reduced power-consumption [110, 111].

A block diagram of a time-interleaved A/D converter is shown in Figure 5.53. In this example, since it uses four sub-A/D converters, it is called a 4-channel or 4-way time-interleave converter. Sampled input values are sent in order from sub-ADC$_1$ to sub-ADC$_4$, and are converted to digital values in each channel. The S/H circuit and switches for distributing and multiplexing the data must be operated at the sampling frequency. On the other hand, it is sufficient for the sub-A/D converters to operate at a rate of 1/4 of the sampling frequency. In other words, the sampling frequency can be quadrupled by using four channels. For example, a 6-bit time-interleaved A/D converter with a sampling frequency of 24 GS/s is reported by using 160 SAR A/D converters [112].

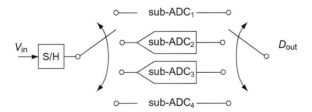

Figure 5.53   Block diagram of a time-interleaved A/D converter.

One of the biggest challenges in time-interleaved A/D converters is to minimize mismatches between the characteristics of each channel. Otherwise, peaks called spurs, frequencies of which are different from the signal frequency, appear in the spectrum. The following describes the effects of mismatches in offset, gain, and sampling-time on the spectrum. The mismatch is exaggerated to show the effect clearly. Actual mismatches are much smaller than these, but the nature of spurs does not change.

Figure 5.54 shows the effect of the offset mismatch in a 4-channel time-interleaved A/D converter. The input frequency of a sinusoidal wave is $f_{in} = 1/(2\pi)$. The outputs from the first channel to the fourth channel are indicated by ◇, □, △, ×, respectively. A sampling frequency is assumed to be 4. If there is an offset of 4 LSB in the fourth channel, quantization errors related to this channel is larger than those related to the other channels. It can then be interpreted that the sinusoidal wave with a period of $1(= 4/4)$

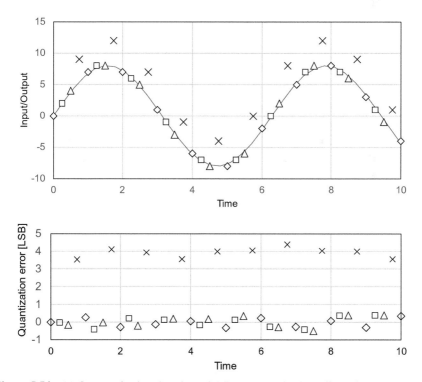

Figure 5.54 (a) Output of a time-interleaved A/D converter having offset mismatch and (b) quantization error.

and its harmonics are generated in addition to the original signal. Therefore, in the spectrum, spurs appear at 1 and its integer multiples. In general, spurs appear at all integer multiples of f_s/N in an N-channel time-interleaved A/D converter.

Figure 5.55 shows the influence of gain mismatch in a 4-channel time-interleaved A/D converter. It is assumed that the gain of the fourth channel is 1.5 times that of the other channels. As with offset, the output from each channel is indicated by $\diamond$, $\square$, $\triangle$, $\times$. In this case, as the input signal amplitude increases, the quantization error also increases. The quantization error is similar to the error due to offset mismatch, but in this case, it can be regarded as amplitude modulated with a frequency of f_{in}. Thus, spurs appear at frequencies that differ by $\pm f_{in}$ from all integer multiples of $f_s/4$.

The influence of sampling-time mismatch in a 4-channel time-interleaved A/D converter is shown in Figure 5.56. It is assumed that the sampling

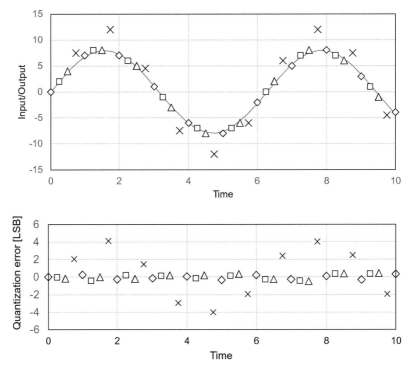

Figure 5.55   (a) Output of a time-interleaved A/D converter having gain mismatch and (b) quantization error.

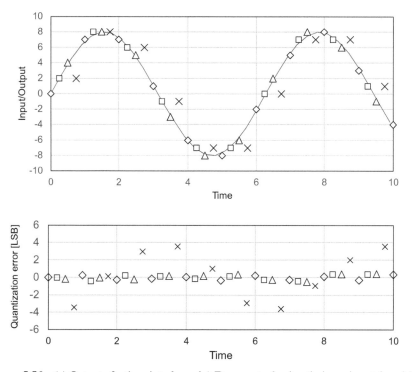

Figure 5.56 (a) Output of a time-interleaved A/D converter having timing mismatch and (b) quantization error.

time of the fourth channel is advanced by two clocks compared with the other channels. The symbols in the figure are the same as before. The quantization error is similar to that observed for the gain mismatch, and in the frequency domain, spurs appear at frequencies different by $\pm f_{in}$ from all integer multiples of $f_s/4$. When the signal changes rapidly, the influence on the error due to the timing mismatch becomes large, so that the error becomes large when the sinusoidal wave crosses zero. Thus, the phase is shifted by $\pi/2$ compared to the gain error case[13].

In summary, the output spectrum in the presence of offset, gain, and sampling time mismatches is shown in Figure 5.57 [29].

Since spurs can be regarded as a noise component, SNR deteriorates as they increase. A schematic diagram of the SNR in time-interleaved A/D converter as a function of the input frequency is shown in Figure 5.58 [113, 114].

---

[13]This corresponds to the phase modulation.

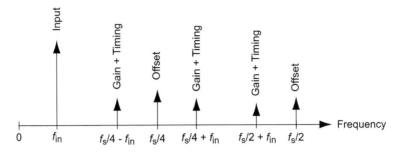

Figure 5.57   Spectrum of a 4-channel time-interleaved A/D converter.

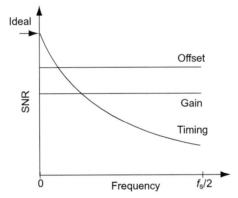

Figure 5.58   Input frequency dependence of SNR in a time-interleaved A/D converter.

Gain mismatch and sampling-time mismatch result in peaks at the same frequencies, but the dependence on the input frequency is different. The former is constant as long as the gain is constant, but in the latter case the peak intensity increases with the input frequency, and the SNR deteriorates. By using this, it is possible to distinguish between them. Although not described here, mismatches in dynamic characteristics in each channel can also cause spurs.

6

Oversampling Analog/Digital (A/D) Converters

Oversampling A/D converters operate at sampling frequencies which are more than ten times higher than the Nyquist rate. Particular attention is paid here to those based on the $\Delta\Sigma$ modulation. First, the basic concept is explained by comparing a $\Delta\Sigma$ modulator with a Nyquist-rate A/D converter. Next, the architectures and characteristics of commonly used $\Delta\Sigma$ modulators are explained. These include first-order and second-order $\Delta\Sigma$ modulators and multi-stage, multi-bit, and continuous-time $\Delta\Sigma$ modulators. Furthermore, decimation filters necessary for constructing $\Delta\Sigma$-type A/D converters are also briefly described. Finally, a D/A converter based on the $\Delta\Sigma$ modulator is presented. Readers who want to know more about these topics may refer to the books dedicated to oversampling converters listed in Section 1.4.

The incremental A/D converter, which has drawn attention in sensor applications recently, is based on the $\Delta\Sigma$-type A/D converter, but the description is omitted in this book. Also, bandpass and quadrature $\Delta\Sigma$ modulators are not explained. Interested readers should refer to the books described in Section 1.4 and a review article on the incremental A/D converter [115].

6.1 Basic Concepts

As described in Chapter 5, various A/D converter architectures have been studied to achieve high resolution. In many cases, the upper limit on the resolution is determined by nonidealities associated with analog circuits, such as nonlinearity of circuits and mismatch in characteristics. Oversampling A/D converters have been developed to break the limit by using sampling frequencies much higher than the Nyquist rate and also using digital signal processing techniques. In other words, the imperfections of analog circuits are fixed with high-speed sampling and digital signal processing. It becomes

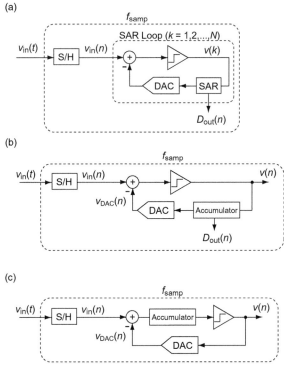

Figure 6.1 Block diagrams of (a) successive approximation (SAR) type A/D converter, (b) Δ modulator, and (c) ΔΣ modulator.

one of the most widely-used A/D converter architectures in recent years due to its compatibility with scaled-down CMOS technology.

The basic concept of an oversampling A/D converter is shown in Figure 6.1 in comparison with that of the SAR A/D converter representing Nyquist-rate A/D converters. Figure 6.1(a) shows a block diagram of a successive-approximation A/D converter, which is basically the same as shown in Figure 5.18. By repeating the comparison, the output value of the internal D/A converter approaches the analog input $v_{in}(t)$. After N iterations, the N-bit output $D_{out}(n)$ is obtained. The input $v_{in}(n)$ sampled at a frequency of f_{samp} corresponds to the digital output $D_{out}(n)$ on one-to-one basis[1].

---

[1]In previous chapters, signals in the time domain were expressed in an uppercase letter such as $V_{in}(t)$. In this chapter, however, according to the convention, they are denoted by a lowercase letter as $v(t)$ to distinguish it from z-transformed signals which are represented using an uppercase letter like $V(z)$.

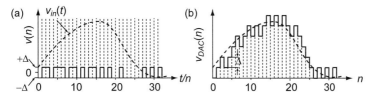

Figure 6.2 (a) Comparator output and (b) internal DAC output in a Δ modulator.

Figure 6.1(b) shows a block diagram of a Δ modulator as a simple example of oversampling A/D converters. Unlike the SAR A/D converter shown in Figure 6.1(a), there is no repetition of the feedback loop for one sample. Instead, every time the input is sampled, the comparator output is accumulated, and the result is used as the feedback signal to the sampled input. An example of the input waveform (v_{in}), the corresponding comparator output ($v(n)$), and the D/A-converter output ($v_{DAC}(n)$) are shown in Figure 6.2. The D/A converter output $v_{DAC}(n)$ follows the analog input with the step height Δ. Here, the comparator output $v(n)$ is assumed to be $\pm\Delta$. When the input increases, the output $+\Delta$ appears more frequently than $-\Delta$. On the other hand, when the input decreases, the output $-\Delta$ appears more frequently than $+\Delta$. The steeper the slope, the higher the frequency. When the input is almost constant, Δ and $-\Delta$ appear almost alternately. The architecture is called a Δ modulator because of the difference taken between the input $v_{in}(n)$ and the feedback signal $v_{DAC}(n)$.

Let us consider how to improve the resolution of the Δ modulation. As Δ becomes small, the step height decreases, which seems useful to obtain higher resolution. However, if the step height is small, $v_{DAC}(n)$ cannot follow an abrupt change in the input signal. For obtaining fast-tracking, it is necessary to increase the sampling frequency. In Figure 6.2, this corresponds to decreasing the step width instead of the step height. Then, even if Δ is kept constant, conversion errors can be reduced by averaging a large number of output values. Since the output $v(n)$ is a digital value, high precision averaging using a digital filter is possible. In other words, by using both high sampling rate and digital processing, resolution can be much improved. This is the basic idea of oversampling A/D converters, details of which will be described below.

Note that in the oversampling A/D converter, unlike the Nyquist type, each sampled input value does not correspond one-to-one with the output value. By averaging oversampled output signals and then decimating them, a high-resolution Nyquist rate output can be obtained. The ratio of the sampling frequency to the Nyquist rate for the input signal is called the oversampling ratio (OSR) and typically ranging from 10 to 1000.

As shown in Figure 6.1(b), the multi-bit D/A converter following the accumulator is included in the feedback path. If the D/A converter has the nonlinearity described in Chapter 4, the conversion performance deteriorates. This is because the output of the D/A converter is subtracted directly from the input signal so that if there is an error due to the nonlinearity, it cannot be distinguished the error from the change in the input signal. A solution to this problem is to use the $\Delta\Sigma$ modulator shown in Figure 6.1(c). In this figure, the feedback path includes a D/A converter, but the resolution required for the D/A converter is different from that in Figure 6.1(b). In the $\Delta\Sigma$ modulator, the input to the D/A converter is the comparator output, which is 1 bit and inherently linear; the line connecting the two points is always a straight line. For this reason, $\Delta\Sigma$ modulators are widely used today[2].

The $\Delta\Sigma$ modulator shown in Figure 6.1(c) can be constructed as follows: Add an accumulator to the input path of the Δ modulator in Figure 6.1(b) and merge it with the accumulator in the loop by moving them just before the comparator. In contrast to the Δ modulator, the output of which represents the difference between successively sampled input signals, the output of the $\Delta\Sigma$ modulator corresponds to the input signal itself. This will be explained in detail in the next sections. Σ in the $\Delta\Sigma$ modulator means that the input signal, as well as the feedback signal, is accumulated[3].

Figure 6.3 shows a typical oversampling A/D converter consists of the $\Delta\Sigma$ modulator described above and a decimation filter [116]. The analog input signal $v_{in}(t)$ is modulated to the N-bit signal $v(n)$ by the $\Delta\Sigma$ modulator. In many cases, the modulator output is 1 bit ($N = 1$). This is because of the inherent linearity as described above. The sampling frequency f_s is equal to the Nyquist frequency f_{snyq} multiplied by the OSR. The OSR is often chosen to be a power of two because it makes the design of the decimation filter easy. The decimation filter down-samples the modulator output $v(n)$ to the Nyquist rate, and at the same time, expands the bit width to

---

[2]A $\Delta\Sigma$ modulator may also use a multi-bit signal in the feedback path. In that case, there are concerns about nonlinearity similar to that in Δ modulators. Details are explained in Section 6.5.

[3]See the column at the end of the chapter for this naming.

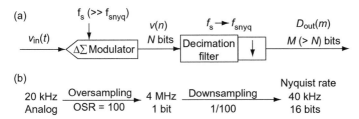

Figure 6.3 (a) Oversampling ADC configuration and (b) numerical example.

M bits[4]. Furthermore, in the decimation filter, it is necessary to sufficiently attenuate the high-frequency signal components that may be folded down to the signal band in the down-sampling process. This will be described in Section 6.7.

In oversampling A/D converters, the quantization noise in the signal band can be reduced, improving the bit resolution. As mentioned in Section 2.2, the quantum noise is assumed to be white noise. When sampling an input signal with signal bandwidth f_B at Nyquist rate $f_{s1}(= 2f_B)$, the noise power is uniformly distributed in the frequency range from dc to $f_{s1}/2$ as shown in Figure 6.4(a). If, for example, the sampling frequency is set to $f_{s2}(= 4f_{s1})$, the noise power spreads to a frequency region which is four times wider as shown in Figure 6.4(b). Since the total noise power is constant, the noise power within the signal band decreases to 1/4. In general, when the sampling frequency is equal to the Nyquist rate multiplied by the OSR, the noise power becomes 1/OSR. Therefore, the SNR is given by

$$\text{SNR (in dB)} = 10 \log \frac{S}{N/\text{SNR}} = 10 \log \frac{S}{N} + 10 \log \text{SNR}. \quad (6.1)$$

Here, S and N are the signal power and the noise power, respectively. This equation shows that doubling the sampling frequency improves the SNR by 3 dB. It corresponds to an improvement of 0.5 bit if represented in the effective number of bits (ENOB).

Furthermore, the quantization noise in the low-frequency region can be moved to the high-frequency region by the $\Delta\Sigma$ modulator used in oversampling A/D converters as shown in Figure 6.4(c). This is called noise shaping. Details will be explained in the following sections, but as shown in

---

[4]This bit width is a nominal value. As will be described later, the bit resolution of this type of A/D converters and modulators is represented by the signal-to-noise ratio (SNR) obtained from the output spectrum or the effective number of bits (ENOB) derived from it.

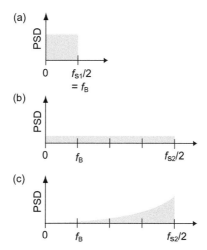

Figure 6.4   Power spectral density (PSD) of quantization noise. (a) Nyquist type ADC, (b) oversampling ADC and (c) noise shaping ADC.

Figure 6.1(c), by paying attention to the accumulator used in the feedback loop, it is possible to understand the reason intuitively. In an ideal negative feedback system in which the loop gain is infinitely large, the feedback signal well follows the input signal. The $\Delta\Sigma$ modulator is also a negative feedback system, and the accumulator contained therein is a lowpass filter. Therefore, it is reasonable that the output can accurately reflect the input in the low-frequency region where the gain is large. Thus, the quantization noise can be suppressed in this region. On the other hand, since the gain is small in the high-frequency region, the difference between the input and the output becomes large, which means that the quantization noise becomes large. This is the noise shaping characteristics shown in Figure 6.4(c).

The oversampling A/D converters have another advantage that the requirement for the anti-aliasing filter characteristic can be relaxed as compared with the Nyquist-rate A/D converters, as is shown in Figure 6.5. In the figure, f_B is the signal bandwidth, and f_{s1} and $f_{s2}(= 4f_{s1})$ are the sampling frequencies under the Nyquist and oversampling conditions, respectively. In Figure 6.5(a), since the signal components, which could be folded into the signal band, exist close to the original signal, a sharp cutoff characteristic is required for the anti-aliasing filter to prevent them from mixing. On the other hand, in the case of the oversampling shown in Figure 6.5(b), since the nearest signal, which could be folded into the signal band, exists rather far away, steep cutoff characteristics are not needed. Therefore, the burden of filter design is significantly reduced.

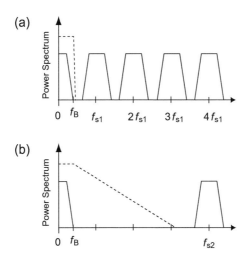

Figure 6.5 Characteristics of anti-aliasing filter required for (a) Nyquist type and (b) over-sampling type.

6.2 1st-order ΔΣ Modulators

The first-order $\Delta\Sigma$ modulator is the simplest one used in oversampling A/D converters, block diagrams of which are shown in Figure 6.6. For explanation, Figure 6.1(c) is shown again as Figure 6.6(a). In Figure 6.6(b), the accumulator is represented with a delay element. Hereafter, according to the convention, the accumulator is called an integrator. In this figure, the linearized model for the comparator (see Section 2.2) is also used. Although the quantization error $e(n)$ depends on the input signal, $e(n)$ is assumed to be a random variable in this model. Note that $y(n)$ in the negative feedback signal and $y(n)$ in the integrator cancel each other. From Figure 6.6(b) the output is represented as

$$v(n) = u(n-1) + e(n) - e(n-1). \tag{6.2}$$

Thus, the quantization error in the first-order $\Delta\Sigma$ modulator is the difference $e(n) - e(n-1)$. In Figure 6.6(c), the position of the delay element in the integrator is changed, and in this case

$$v(n) = u(n) + e(n) - e(n-1) \tag{6.3}$$

is satisfied.

For intuitive understanding, the signal waveforms of each node for a dc input of 0.3 are shown in Figure 6.7. The output $v(n)$ is the quantized output

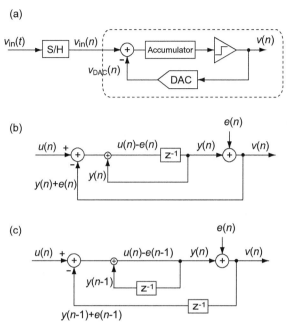

Figure 6.6   (a) Block diagram of a first-order $\Delta\Sigma$ modulator, the same one as shows in Figure 6.1(c), (b) linearized diagram, and (c) modified one. To simplify the following description, $v_{\mathrm{in}}(n)$ in (a) is rewritten as $u(n)$ in (b) and (c).

value, and as n increases, the averaged output gradually approaches the input value 0.3 as shown in Figure 6.7(c). The input $y(n)$ to the comparator satisfies

$$y(n) = u(n-1) - e(n-1) = y(n-1) + u(n-1) - v(n-1)$$
$$= y(0) + \sum_{k=0}^{n-1} (u(k) - v(k)). \tag{6.4}$$

By rearranging the terms above, the average of $u(n)$ can be expressed as

$$\bar{u} = \lim_{N\to\infty} \frac{1}{N} \sum_{k=0}^{N} u(k) = \lim_{N\to\infty} \frac{1}{N} \left[y(n+1) - y(0) + \sum_{k=0}^{N} v(k) \right]$$
$$= \lim_{N\to\infty} \frac{1}{N} \sum_{k=0}^{N} v(k). \tag{6.5}$$

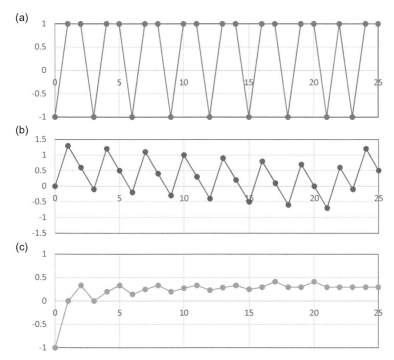

Figure 6.7 Signal waveforms of each node for a dc input (= 0.3). (a) Output $v(n)$, (b) input to comparator $y(n)$, and (c) cumulative average of output $v(n)$.

In the deviation above, it is used that $(y(n) - y(0))/N$ becomes 0 as N becomes infinite because y (n) is bounded. In other words, if the number of output values to be averaged is sufficiently large, it is possible to digitally represent the input value with any degree of accuracy at least in principle.

Evaluation of stability is essential in feedback systems. When the feedback signal from the comparator is ± 1, the necessary condition for stable operation is $|u(n)| < 1$. If $|u(n)| > 1$, it does not work correctly. For example, if $u(n) = 1.1$, the integrator output will continue to increase even if the feedback signal is -1 constantly. Also, when the input is represented by a simple integer ratio such as 1/2, the output changes periodically, which is called an idle tone.

Figure 6.8 shows the simulated waveforms of the first-order $\Delta\Sigma$ modulator for a sinusoidal input. Here, the digital output of the $\Delta\Sigma$ modulator is ± 0.5 V. As can be inferred from the result for dc inputs, when the input signal is large, the frequency of appearance of 0.5 V increases. In contrast, when the input signal is small, the appearance frequency of -0.5 V increases.

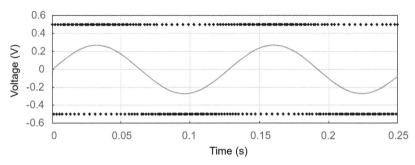

Figure 6.8   First-order $\Delta\Sigma$ modulator output.

Now, let us estimate the noise power spectral density of the first-order $\Delta\Sigma$ modulator based on the linear model. According to Figure 6.6(b)

$$v(n) = u(n-1) + e(n) - e(n-1). \tag{6.6}$$

The z transform results in

$$V(z) = z^{-1}U(z) + (1 - z^{-1})E(z). \tag{6.7}$$

Generally, the output is expressed as

$$V(z) = \text{STF}(z)U(z) + \text{NTF}(z)E(z), \tag{6.8}$$

where STF and NTF are the signal transfer function and the noise transfer function, respectively. By comparing these equations,

$$\text{NTF}(z) = 1 - z^{-1} \tag{6.9}$$

is obtained as a noise transfer function. Replacing z with $\exp\left(j2\pi fT_{\text{s}}\right)$, the NTF in the frequency domain is obtained as

$$|\text{NTF}(f)| = |1 - \exp\left(-j2\pi fT_{\text{s}}\right)| = 2\sin\pi fT_{\text{s}} \cong 2\pi fT_{\text{s}}. \tag{6.10}$$

This equation means that the noise power increases with the frequency at 20 dB/dec in the low-frequency region. This is referred to as the noise shaping.

If the total power of quantization noise is e_{rms}^2, the spectral density $S_{\text{e}}(f)$ can be written as

$$S_{\text{e}}(f) = \frac{e_{\text{rms}}^2}{f_{\text{s}}/2}. \tag{6.11}$$

Therefore, the noise power in the signal band $[0, f_B]$ is

$$\int_0^{f_B} (2\sin\pi f T_s)^2 \frac{2e_{rms}^2}{f_s} df \cong \int_0^{f_B} (2\pi f T_s)^2 \frac{2e_{rms}^2}{f_s} df$$

$$= \frac{\pi^2 e_{rms}^2}{3\,(\text{OSR})^3}. \qquad (6.12)$$

This expression shows that the signal-to-noise ratio (SNR) can be improved by 9 dB for doubling the OSR. As explained in the previous section, doubling the OSR results in an improvement of 3 dB without noise shaping. Thus the effect of noise shaping is significant. The simulation result of the spectrum for a sine-wave input is shown in Figure 6.9. The peak around 1 Hz represents an input wave. It can be confirmed that the noise power increases with 20 dB/dec, which agree with the discussion above.

Note that the analysis described above is based on the linear model shown in Figure 6.6(b), where the nonlinear comparator quantization noise is replaced with random quantization noise. Since the input signal uniquely determines the quantization noise in really, the linearized model is an approximation for convenience. On the other hand, the simulation result shown in Figure 6.9 includes the nonlinearity. Since the slope of the noise power spectral density almost agrees with the predicted value of 20 dB/dec from the linear model, the linear model is sufficiently valid in this sense. In general, however, this model may not hold in some cases, for example, where the ratio of the input signal frequency to the sampling frequency is a simple

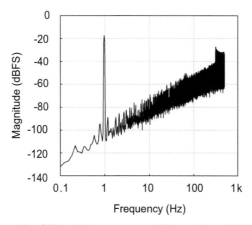

Figure 6.9 First-order ΔΣ modulator spectrum. The number of FFT points is 16384.

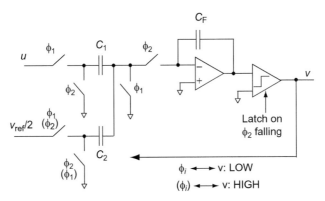

Figure 6.10   First-order $\Delta\Sigma$ modulator circuit.

integer ratio. It should be emphasized that simulation reflecting nonlinearity is indispensable for actual circuit designs.

A circuit schematic of a first-order $\Delta\Sigma$ modulator is shown in Figure 6.10. The circuit operates as follows. First, let us assume that $v(n)$ is LOW. As shown in Figure 6.11, when ϕ_1 is HIGH, the circuit is in the sample mode, and C_1 and C_2 are charged with $u(n)$ and $v_{\text{ref}}/2$, respectively. When ϕ_2 is HIGH, the circuit is in the integration mode. Since the inverting terminal of the opamp is virtually grounded, all the charges of C_1 and C_2 are transferred to C_F. This results in

$$v_1 = v_0 + \frac{C_1}{C_F}u(n) + \frac{C_2}{2C_F}v_{\text{ref}}. \tag{6.13}$$

Similarly, when $v(n)$ is HIGH, it becomes

$$v_1 = v_0 + \frac{C_1}{C_F}u(n) - \frac{C_2}{2C_F}v_{\text{ref}}. \tag{6.14}$$

These equations show that the negative feedback value is subtracted from the input, the result of which is added to the previously accumulated value of v_0.

In a practical implementation, it is necessary to consider that the opamp gain A in the integrator is finite. It is known that the influence on noise transfer function can be neglected when $A >$ OSR. If there is "leakage" in the integrator, and if the absolute value of the input is less than $(1/2)A$, the integration will not be executed and the input will not be reflected in the output even if the input changes. This is called a dead zone. In either case, detailed circuit simulation is indispensable for predicting circuit performance.

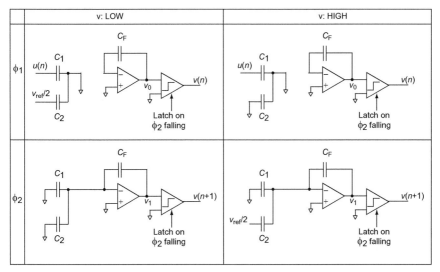

Figure 6.11 Operation of the circuit shown in Figure 6.10.

This is the case not only for the first-order $\Delta\Sigma$ modulator but also for $\Delta\Sigma$ modulator circuit designs in general. Detailed explanations go beyond the scope of this book, but readers interested should refer to the literature listed at the beginning of this chapter. Also, note that the correlated double sampling [117] is known as a method to eliminate the influence of the finite gain of opamps.

6.3 2nd-order $\Delta\Sigma$ Modulators

A block diagram of the second-order $\Delta\Sigma$ modulator is shown in Figure 6.12. By incorporating the first-order $\Delta\Sigma$ modulator as a quantizer into another first-order $\Delta\Sigma$ modulator as shown in Figure 6.12(a), the second-order $\Delta\Sigma$ modulator is obtained. From this figure, the output $v(n)$ is derived as

$$v(n) = u(n-1) + E(n) - E(n-1)$$
$$= u(n-1) + (e(n) - e(n-1)) - (e(n-1) - e(n-2)). \quad (6.15)$$

The last two terms represent the second-order noise shaping. Figure 6.12(b) is a representation using delay elements. This is obtained by rearranging delay elements in Figure 6.12(a).

Figure 6.13 shows waveforms of each node against a dc input of 0.3. At first glance, there is no significant difference from the case of the first-order

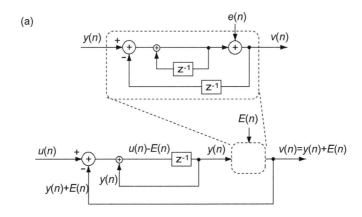

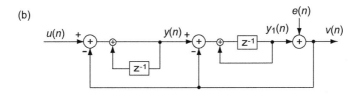

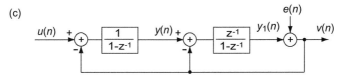

Figure 6.12   (a) Construction of a second-order $\Delta\Sigma$ modulator based on first-order $\Delta\Sigma$ modulators. Block diagrams represented by using (b) delay elements and (c) transfer functions.

$\Delta\Sigma$ modulator, and the averaged value approaches 0.3 asymptotically. The simulation result of the output for the sine-wave input is shown in Figure 6.14. It can be seen that the appearance frequency of HIGH and LOW corresponds to the input waveform, and this looks also similar to the first-order $\Delta\Sigma$ modulator.

The difference from the first-order $\Delta\Sigma$ modulator certainly appears in the output spectrum. By performing the z transform

$$
\begin{aligned}
V(z) &= z^{-1}U(z) + (1 - z^{-1})E(z) - (z^{-1} - z^{-2})E(z) \\
&= z^{-1}U(z) + (1 - z^{-1})^2 E(z)
\end{aligned}
\tag{6.16}
$$

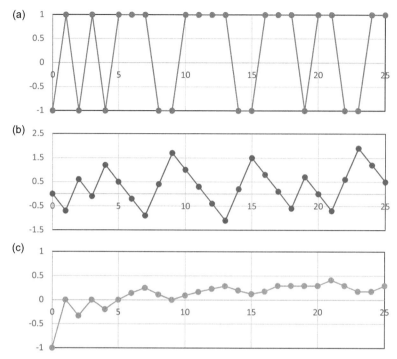

Figure 6.13 Signal waveforms of each node for a dc input (= 0.3). (a) Output $v(n)$, (b) input to comparator $y(n)$, and (c) cumulative average of output $v(n)$.

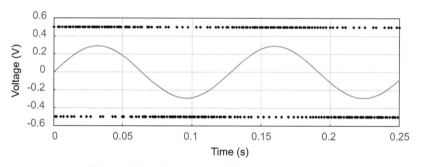

Figure 6.14 Second-order $\Delta\Sigma$ modulator output.

can be derived from Equation (6.15). $(1 - z^{-1})^2$ is the noise transfer function of the second-order $\Delta\Sigma$ modulator. The NTF in the frequency domain is obtained by replacing z with $\exp(j2\pi f T_s)$, resulting in

$$|\text{NTF}(f)| = |1 - \exp(-j2\pi f T_s)|^2 = (2\sin\pi f T_s)^2 \cong (2\pi f T_s)^2. \quad (6.17)$$

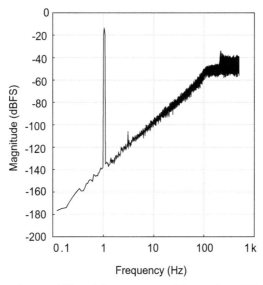

Frequency (Hz)

Figure 6.15   Second-order $\Delta\Sigma$ modulator spectrum. The number of FFT points is 16384.

Therefore, in the low-frequency region, the noise power increases with the frequency with a slope of 40 dB/dec. It is twice the value in the first-order $\Delta\Sigma$ modulator. The simulation result of the second-order $\Delta\Sigma$ modulator spectrum is shown in Figure 6.15. The slope of 40 dB/dec is consistent with the prediction of the linear model described above.

Figure 6.16 indicates the effective number of bits (ENOB) as a function of the input amplitude. Since the quantization noise power is independent of the input amplitude, the SNR increases as the amplitude increases with a slope of 20 dB/dec. Then ENOB increases by approximately 3.3 bits/dec according to Equation (2.47). The slope in the simulation result almost agrees with this value. If the plot is extrapolated to the smaller input amplitude, the input amplitude at which ENOB becomes 0 can be obtained. Since a dynamic range (DR) can be obtained from this extrapolation, the graph is called a DR plot.

By the same procedure as was explained in the first-order $\Delta\Sigma$ modulator, the noise power in the signal band is calculated as

$$\int_0^{f_B} (2\sin \pi f T_s)^4 \frac{2e_{\mathrm{rms}}^2}{f_s} df \cong \int_0^{f_B} (2\pi f T_s)^4 \frac{2e_{\mathrm{rms}}^2}{f_s} df$$

$$= \frac{\pi^2 e_{\mathrm{rms}}^2}{5\,(\mathrm{OSR})^5}. \tag{6.18}$$

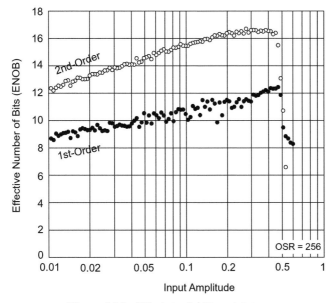

Figure 6.16 DR plots of ΔΣ modulators

This expression means that if the OSR is doubled, the signal-to-noise ratio (SNR) can be improved by 15 dB. The relationship between the oversampling ratio (OSR) and effective resolution (ENOB) is shown in Figure 6.17. The noise power in the signal band can be obtained in the same manner used for the first-order ΔΣ modulator. The ENOB increases as the OSR increases with 1.5 bit/oct for the first-order ΔΣ modulator and 2.5 bit/oct for the second-order ΔΣ modulator. These values are in agreement with the values obtained from the linear model.

A circuit diagram of the second-order ΔΣ modulator is shown in Figure 6.18. This circuit adopts integrators using an inverter instead of a conventional opamp to reduce power consumption. The operation of the inverter-based integrator is explained by using Figure 6.19. Figure 6.19(b) shows the circuit diagram in the sampling mode where ϕ_1 is closed, and the charge $-C_S v_{in}$ is stored in the right-side electrode of C_S. Also, the charge $C_C v_1$ is stored in the left-side electrode of C_C. v_1 is the logical threshold voltage, which is the voltage when the input and output of the inverter are the same. Figure 6.19(c) shows the circuit in the integration mode, where ϕ_2 is closed. Since the right-side electrode of C_C is connected to the inverter's input, which is a high impedance node, the charge stored in C_C does not

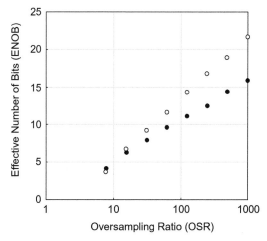

Figure 6.17   Bit resolution (ENOB) as a function of OSR for first-order (•) and second-order (○) $\Delta\Sigma$ modulators.

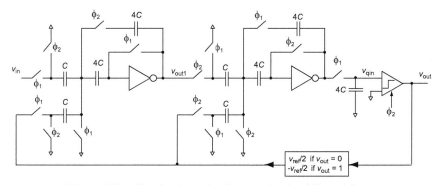

Figure 6.18   Circuit schematic of a second-order $\Delta\Sigma$ modulator.

change. Thus, the potential difference across C_C does not change either. Also, if the gain of the inverter is assumed to be sufficiently large, the voltage of the inverter input terminal is v_1 because of the negative feedback. Therefore, $v_x = 0$, and the charge stored in the right-side electrode of C_S ($-C_S v_{in}$) is transferred to the left-side electrode of C_F. Then, this charge is added to the charge previously stored in C_F, and an integral function is realized.

Intensive discussion on the practical design criteria for implementing second-order modulators by including various circuit impairments is presented in the literature [118].

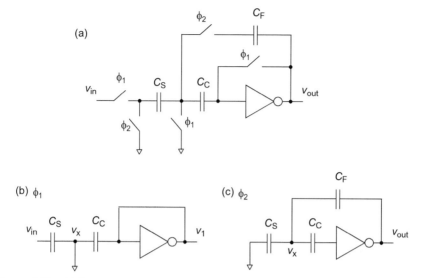

Figure 6.19 (a) Inverter-based integrator. (b) and (c) indicate the sampling mode and the integration mode.

6.4 Multi-stage ΔΣ Modulators

Compared with the first-order ΔΣ modulator, the second-order ΔΣ modulator is more suitable for high-resolution operation. It is then expected that resolution will be still further improved if a higher-order ΔΣ modulator is constructed. However, since the ΔΣ modulator is a feedback system, stability is a major concern in higher-order configurations, as is the case with multi-stage opamp designs. Instead of including a higher-order filter in a single feedback loop, a configuration consisting of several first-order or second-order loops has been proposed, which results in higher-order noise shaping characteristics with excellent stability. In this section, a multi-stage ΔΣ modulator, commonly referred to as MASH (multi-stage noise-shaping) [119], is described.

Figure 6.20 shows a MASH ΔΣ modulator composed of two first-order ΔΣ modulators. $e_1(n)$ is the quantization error in the first stage, which is the input to the second-stage ΔΣ modulator. The two outputs are added after passing through the digital filters to obtain the final output. The output from the first stage $v_1(n)$ is

$$v_1(n) = u(n-1) + (e_1(n) - e_1(n-1)), \tag{6.19}$$

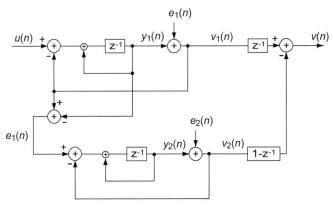

Figure 6.20   MASH $\Delta\Sigma$ modulator composed of two first-order $\Delta\Sigma$ modulators.

and the output from the second stage $v_2(n)$ can be written as

$$v_2(n) = e_1(n-1) + (e_2(n) - e_2(n-1)). \tag{6.20}$$

By adding these using the digital filters shown in the figure,

$$\begin{aligned}
v(n) &= v_1(n-1) - (v_2(n) - v_2(n-1)) \\
&= u(n-2) - (e_2(n) - e_2(n-1)) \\
&\quad + (e_2(n-1) - e_2(n-2))
\end{aligned} \tag{6.21}$$

is obtained. This is the same result as Equation (6.15), indicating that the second-order noise shaping can be realized.

One might ask how to construct the two digital filters. Let us answer that question using Figure 6.21 showing the two-stage MASH scheme redrawn by using two loop filters. From this figure,

$$V_1(z) = L_{10}(z)U(z) + L_{11}(z)V_1(z) + E_1(z) \tag{6.22}$$

is obtained. Solving for $V_1(z)$ yields

$$V_1(z) = \frac{L_{10}}{1 - L_{11}}U(z) + \frac{1}{1 - L_{11}}E_1(z). \tag{6.23}$$

Therefore, the signal transfer function $\mathrm{STF}_1(z)$ and the noise transfer function $\mathrm{NTF}_1(z)$ are obtained as

$$\mathrm{STF}_1(z) = \frac{L_{10}}{1 - L_{11}} \tag{6.24}$$

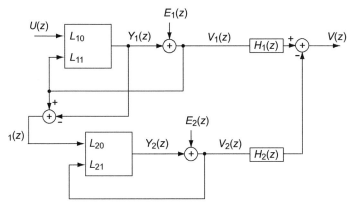

Figure 6.21 MASH ΔΣ modulator represented by loop filters.

$$\mathrm{NTF}_1(z) = \frac{1}{1 - L_{11}}. \tag{6.25}$$

By using these equations, the input/output relation of the two ΔΣ modulators can be written as

$$V_1(z) = \mathrm{STF}_1(z)U(z) + \mathrm{NTF}_1(z)E_1(z) \tag{6.26}$$

$$V_2(z) = \mathrm{STF}_2(z)E_1(z) + \mathrm{NTF}_2(z)E_2(z). \tag{6.27}$$

Therefore, the final output $V(z)$ can be obtained as

$$\begin{aligned}
V(z) &= H_1(z)(\mathrm{STF}_1(z)U(z) + \mathrm{NTF}_1(z)E_1(z)) \\
&\quad - H_2(z)(\mathrm{STF}_2(z)E_1(z) + \mathrm{NTF}_2(z)E_2(z)) \\
&= H_1(z)\mathrm{STF}_1(z)U(z) \\
&\quad + (H_1(z)\mathrm{NTF}_1(z) - H_2(z)\mathrm{STF}_2(z))E_1(z) \\
&\quad - H_2(z)\mathrm{NTF}_2(z)E_2(z).
\end{aligned} \tag{6.28}$$

Thus, if

$$H_1(z)\mathrm{NTF}_1(z) = H_2(z)\mathrm{STF}_2(z), \tag{6.29}$$

$E_1(z)$ can be cancelled. If the signal transfer function is approximated to 1,

$$V(z) = H_1(z)U(z) - H_1(z)\mathrm{NTF}_1(z)\mathrm{NTF}_2(z)E_2(z) \tag{6.30}$$

is obtained. Then, the noise transfer function is $H_1(z)\mathrm{NTF}_1(z)\mathrm{NTF}_2(z)$. This means that the sum of the order of the first stage and the second stage becomes the overall noise-shaping order. If another stage is added, higher-order noise shaping characteristics can be obtained in principle while maintaining the stability of the lower-order $\Delta\Sigma$ modulator.

The MASH scheme has other advantages. Since the quantization noise close to white noise is used as the input of the second stage, it is possible to suppress a tone which is likely to occur for a periodic input. Also, there is no feedback from the second stage to the analog input, and H_2 has highpass characteristics as can be seen from Equation (6.29). Therefore, even if a multi-bit output quantizer is used in the second stage, the influence of its nonlinearity is small.

On the other hand, if the matching of the filters is incomplete, the quantization noise leaks. This can be understood from the fact that if Equation (6.29) is not accurately satisfied, quantization noise of the first stage cannot be completely canceled out. To prevent this, it is necessary to match the characteristics of the digital filters and the analog loop filters. More specifically, it is necessary to match the capacitance and increase the gain in the analog circuit. An alternative approach to solving this problem is to use the sturdy MASH (SMASH) scheme. Those who are interested in it, please refer to the article [120].

6.5 Multi-bit $\Delta\Sigma$ Modulators

As mentioned in Sections 6.2 and 6.3, it is useful to raise the OSR to improve the bit resolution of $\Delta\Sigma$-type A/D converters. This means, however, that it is necessary to operate the circuit at a higher speed. Especially, obtaining a good resolution for high-frequency input signals is a serious challenge, as there is a limit to speeding up the circuits. In other words, the OSR cannot be made sufficiently high for wideband input signals, and there is a limit in the bit resolution. Using the higher-order loop filter is also an effective means of increasing the bit resolution, but stability becomes another major challenge. The multi-bit $\Delta\Sigma$ modulator explained in this section has been studied as an alternative solution.

The block diagram of a 2-bit $\Delta\Sigma$ modulator is shown in Figure 6.22. Figure 6.23 shows the output of a 3-bit $\Delta\Sigma$ modulator obtained by simulation using a behavior model. The output is a value of 3 bits (8 levels). Like the previous 1-bit $\Delta\Sigma$ modulator, the output distribution reflects the magnitude of the input signal.

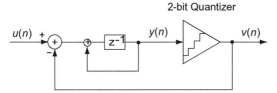

Figure 6.22 Block diagram a 2-bit ΔΣ modulator.

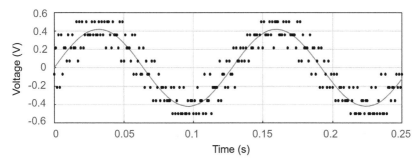

Figure 6.23 3-bit ΔΣ modulator output.

By using a multi-bit signal in the feedback path, the quantization noise power can be reduced. The noise power in the signal band can be written as

$$n_0^2 = \frac{e_{\text{rms}}^2}{\text{OSR}} = \frac{\Delta^2}{12 \cdot \text{OSR}} = \frac{V_{\text{FS}}^2}{12 \cdot \text{OSR} \cdot Q^2}. \tag{6.31}$$

Here, Δ is the quantization interval, V_{FS} is the full-scale voltage, and Q is the number of quantization levels. The above equation shows that the SNR can be improved by 6 dB by doubling Q or increasing the quantizer bit resolution by one. This corresponds to the ENOB being improved by 1 bit. Figure 6.24 shows a simulation result showing improvement in the bit resolution by adopting the multi-bit scheme. It can be confirmed that the ENOB is improved by 1 bit each time the number of bits of the quantizer is increased by one.

In addition, other advantages of using multi-bit modulators are as follows: (1) Design conditions of decimation filter can be relaxed, (2) the reduced quantization noise improves the stability of the modulator, (3) the wide margin in the NTF design makes it possible to optimize the circuit parameters aggressively, and (4) requirements for the opamp slew rate in the loop filter are relaxed.

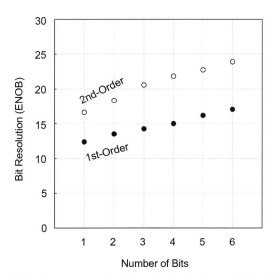

Figure 6.24   Improvement in bit resolution by the multi-bit scheme.

On the other hand, there is an important issue to consider: nonlinearity of the D/A converter used in the feedback path. Since the output of the D/A converter is directly added to the input, if there is nonlinearity, the input signal itself seems to have changed, and the SNR deteriorates. In other words, the nonlinearity is not noise-shaped. One efficient method to reduce the influence of nonlinearity is known as dynamic element matching (DEM) [54]. The principle is shown in Figure 6.25. Let us assume that the output of the quantizer is 3 bits and a 3-bit current-steering D/A converter with seven unit-current sources are used for generating the feedback signal. If the 3-bit output is 101 as shown in this figure, the corresponding analog output is obtained by choosing 5 out of 7 current sources.

Conventionally, the combination of 5 current sources out of 7 is fixed. If there is a mismatch in the current values of each current source, nonlinearity errors occur depending on the current values.[5] On the other hand, the DEM randomly determines the combination of five current sources out of the seven at every feedback cycle. If all the current sources have the same probability of being chosen, the current averaged over a sufficiently long time approaches an accurate value even though the output value at each time is different due to

---

[5]Readers who are interested in variations in the characteristics of MOSFETs, refer to the frequently-cited paper [121].

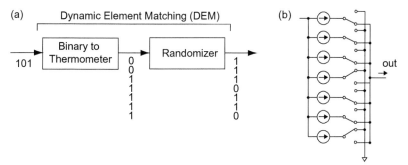

Figure 6.25 (a) Dynamic element matching (DEM) operation and (b) circuit implementation.

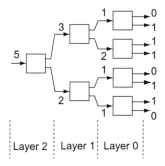

Figure 6.26 Block diagram of a tree type DEM.

the mismatch. The idea of time averaging works well with the oversampling A/D converter, and the DEM works effectively.

Various randomization algorithms are known, such as a tree-type [122] and a data weighted averaging (DWA) [123]. An example of a tree type is shown in Figure 6.26. Three layers of switch box are required for the 3-bit output. The sum of the two outputs of the switch box is equal to its input, and if the input is an even number, the value is divided by 1/2. If the input is odd, given $2n+1$, n and $n+1$ are the two outputs, and they alternate at each sampling. Simulation result when the output varies from 0 to 6 is shown in Figure 6.27. Solid-circled current sources are selected to provide the output current. It can be seen that the way of choosing the current sources is not fixed and they are selected with equal probability. The deterioration of SNR caused by DAC mismatch error and the effect of DEM is shown in Figure 6.28. By adopting the DEM, the tolerance to mismatches is improved.

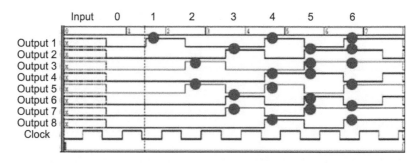

Figure 6.27   Simulation results on DEM

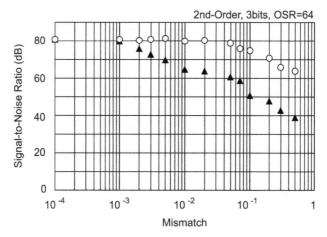

Figure 6.28   Improvement in SNR by introducing DEM.

6.6 Continuous-time $\Delta\Sigma$ Modulators

$\Delta\Sigma$ modulators described in the previous sections were implemented by using switched-capacitor circuits. First, the S/H circuit discretizes the input signal, followed by the discrete-time (DT) loop filter. As described in Section 3.1.2, the sampling period should be sufficiently long compared with the settling time of the circuit, which is typically several times the inverse of the 3-dB frequency of the opamp. For digital audio applications, this is not a significant limitation. In contrast, for wireless communication applications, it is difficult to operate the circuit at high speed to obtain a high OSR. A possible solution is to use a multi-bit $\Delta\Sigma$ modulator explained in the previous section, which is capable of realizing high resolution even at a low OSR. However, the DEM technique is necessary. In this section, as an alternative approach, ΔΣ modulators that operate with continuous-time (CT) circuits are presented. Called continuous-time (CT) ΔΣ modulators, they can operate with a relatively high OSR to accommodate a full bandwidth because they do not have to wait for the CR settling to be completed.

Block diagrams of a continuous-time (CT) ΔΣ modulator is shown in Figure 6.29 with those of a discrete-time (DT) ΔΣ modulator. Figure 6.29(a) shows a block diagram of a DT ΔΣ modulator, which is rewritten as that

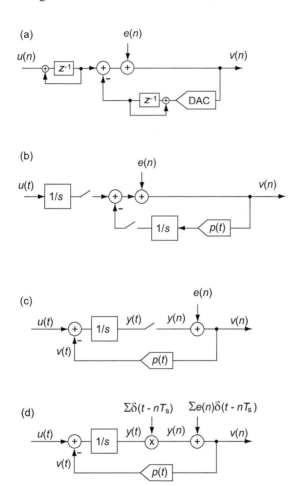

Figure 6.29 (a) Discrete-time (DT) ΔΣ modulator with DT integrators located before the feedback summing node, (b) continuous-time (DT) ΔΣ modulator with CT integrators, (c) CT ΔΣ modulator with an integrator moved back after the summing node, and (d) CT ΔΣ modulator where the switch is expressed by using a multiplier.

shown in Figure 6.29(b). Changing the position of the integrators $(1/s)$ results in a CT $\Delta\Sigma$ modulator shown in Figure 6.29(c). Figure 6.29(d) shows the same one with the switch in Figure 6.29(c) represented by a multiplier. Suppose that $y(n)$ is the value after sampling the continuous-time signal $y(t)$ with a period of T_s. Then, $y(n)$ can be expressed as

$$y(n) = \sum_n y(t)\delta(t - nT_s). \tag{6.32}$$

The discrete-time signal $v(n)$, which is the output of the comparator, can be written as

$$v(n) = \sum_n [y(t) + e(n)]\,\delta(t - nT_s), \tag{6.33}$$

where $e(n)$ is the quantization noise. The continuous-time signal $v(t)$ is

$$v(t) = \sum_n v(n)p(t - nT_s). \tag{6.34}$$

Here, $p(t)$ is a function that defines an output waveform of the feedback signal. $v(t)$ represents an NRZ (non-return to zero) signal as shown in Figure 2.15 if

$$p(t) = \begin{cases} 1 & 0 < t < T_s \\ 0 & \text{(other ise)} \end{cases}. \tag{6.35}$$

Now, consider the impulse response of a continuous-time modulator and a discrete-time modulator using Figure 6.30. By cutting the loop at P in the block diagram of Figure 6.30(a), the block diagram of Figure 6.30(b) is obtained. If the impulse train $y_D(n) = 1, 0, 0, 0, \cdots$ is assumed as the quantization noise, the returning signal $y'_D(n)$ is $0, -1, -1, -1, \cdots$. Similarly, consider Figure 6.30(d) obtained by cutting the loop shown in Figure 6.30(c) at P'. The response $y'_C(n)$ to the impulse sequence $y_C(n) = 1, 0, 0, 0, \cdots$ can be expressed as

$$y'_C(n) = p(t) * h(t). \tag{6.36}$$

Where $*$ means the convolution, and $h(t)$ is the impulse response of the integrator. Assuming that $p(t)$ satisfies

$$\int_0^1 p(t)dt = 1, \tag{6.37}$$

(a)

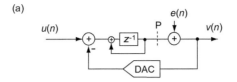

(b)

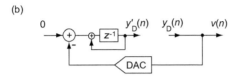

(c)

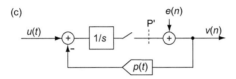

(d)

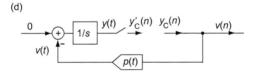

(e)

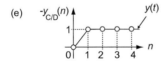

Figure 6.30 (a) DT ΔΣ modulators, (b) that with a loop opened at P, (c) CT ΔΣ modulators, (d) that with a loop opened at P′, and (e) DT (open circles) and CT (solid line) impulse responses.

the returned signal $y'_C(n)$ is $0, -1, -1, -1, \cdots$, which is equal to $y'_D(n)$. Therefore, if only sampling points are taken out, the same noise transfer function (NTF)

$$NTF(z) = 1 - z^{-1} \tag{6.38}$$

as in Figure 6.30(a) is obtained in Figure 6.30(c).

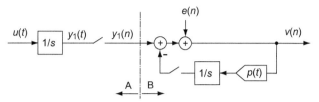

Figure 6.31   Block diagram of a CT $\Delta\Sigma$ modulator with two integrators before the feedback summing node, the same as shown in Figure 6.29(b).

An essential property of the continuous-time $\Delta\Sigma$ modulator is that the anti-aliasing function is included in the signal transfer function. To explain this, let us consider the signal transfer function (STF) by using Figure 6.31, which is the same as Figure 6.29(b). Let the input $u(t)$ be represented by $\exp(j2\pi ft)$. Since the continuous-time signal $y_1(t)$ after passing through the integrator can be written as

$$y_1(t) = \frac{1}{j2\pi f}e^{j2\pi ft}, \tag{6.39}$$

the discrete-time signal $y_1(n)$ after sampling can be expressed as

$$y_1(n) = \frac{1}{j2\pi f}e^{j2\pi fn}. \tag{6.40}$$

Here, the sampling period T_s is assumed to be 1. When this signal enters the feedback loop, the output can be written as

$$v(n) = y_1(n) + e(n) - y_1(n-1) - e(n-1) \tag{6.41}$$

$$= \frac{1}{j2\pi f}(1 - e^{-j2\pi f})e^{j2\pi fn} + e(n) - e(n-1). \tag{6.42}$$

Therefore, the signal transfer function $\mathrm{STF}(f)$ is obtained as

$$\mathrm{STF}(f) = \frac{1}{j2\pi f}(1 - e^{-j2\pi f}) = e^{-j\pi f}\frac{\sin \pi f}{\pi f}. \tag{6.43}$$

Since the sinc function on the righthand side of the above equation becomes 0 when $f = 1, 2, 3, \cdots$, $\mathrm{STF}(f)$ also becomes 0 at these frequencies. This means that signal components existing around $f = 1, 2, 3, \cdots$, which would be folded back to the low-frequency signal band by sampling, can be eliminated by $\mathrm{STF}(f)$. In other words, in the continuous-time $\Delta\Sigma$ modulator, an anti-aliasing filter is built in as a property of $\mathrm{STF}(f)$[6].

---

[6] An example in which high-speed operation and high-energy efficiency are achieved with anti-aliasing function in a continuous-time pipelined A/D converter was reported [124].

This can be intuitively understood by looking at Figure 6.31. Even if the signal having the sampling frequency $f_s(= 1)$ or a frequency that is an integral multiple of the sampling frequency is included in $u(t)$, it disappears after being integrated over the time $1/f_s$, and it is not included in $y_1(t)$. This is the same situation that the integral-type A/D converter described in Section 5.7.1 can remove the noise having the frequency component being the same as the reciprocal of the integration time.

In this way, the continuous-time ΔΣ modulator is excellent in high-speed operation and inherently has the anti-aliasing characteristics so that it becomes popular in wireless communication applications [125]. However, there are points to be noted: Those are the excess loop delay and the jitter tolerance which are explained below.

The effect of a loop delay is shown in Figure 6.32. The influence of a delay τ appears in $y'_c(n)$ as shown in Figure 6.32(b). Figure 6.33 shows a technique to compensate for the loop delay [126], which restores the delayed response to the ideal one shown in Figure 6.30(e). As shown in the figure, adding a new route to compensate for the attenuation due to the delay can restore the ideal response shown in Figure 6.30(e).

Figure 6.34 shows the response when there is time jitter in the switch in the feedback path shown in Figure 6.31. In an actual circuit, it means that jitter is included in the clock applied to the D/A converter in the feedback path. The effect of jitter can be modeled by varying the pulse width of $p(t)$. Assuming that the sampling period is T_s and the standard deviation of jitter is σ_j, it is known that the signal-to-noise ratio SNR_{jitter} caused by the jitter can

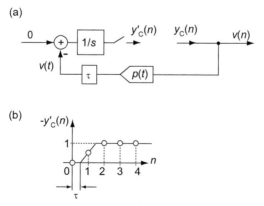

Figure 6.32 (a) Block diagram including a loop delay of τ and (b) resulting impulse response.

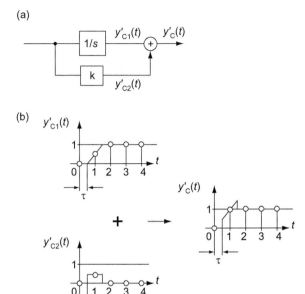

Figure 6.33   (a) Block diagram with loop delay compensation and (b) improvement in the impulse response.

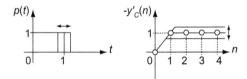

Figure 6.34   (a) Pulse width variation of $p(t)$ due to jitter and (b) resulting impulse response.

be written as [127]

$$\text{SNR}_{\text{jitter}} = 10 \, \log_{10} \left(\frac{T_s^2 \cdot \text{OSR}}{\sigma_j^2} \right). \tag{6.44}$$

In order to reduce the effect of jitter, it is useful to modify the waveform of $p(t)$ while maintaining the normalization condition of Equation (6.37). It is sufficient to set $p(t)$ to 0 around $t = 1$ and to increase $p(t)$ in other region to satisfy Equation (6.37). This is known as feedback pulse shaping for DAC signals: for example, adopting RTZ (return-to-zero) pulses [128], generating pulses with SC-R circuit [129], waveform shaping with FIR filter [130], sine-shaped pulse [131], pulsing by delay processing [132], and averaging jitter by DACs connected in parallel [133]. Detailed descriptions are beyond the scope of this book, but interested readers should refer to the cited documents above, as well as tutorial reviews on CT $\Delta\Sigma$ architecture [26, 134, 135].

6.7 Decimation Filters

This section explains decimation filters, which are required together with a $\Delta\Sigma$ modulator to construct a $\Delta\Sigma$-type A/D converter. Its role is to down-sample the oversampled $\Delta\Sigma$ modulator output to a Nyquist-rate signal. After the downsampling, the number of $\Delta\Sigma$ modulator output bits (usually 1 bit) increases to the final number of output bits of the A/D converter. In addition to changing the sampling rate, it is necessary to prevent the folding back of the high-frequency components into the signal band in the downsampling. Digital comb filters are suitable for this purpose.

Since a typical decimation ratio, which is the reciprocal of the OSR, is a value of tens to hundreds, if it is attempted to process it with a single stage, the filter configuration becomes so complicated that not only the power consumption increases but also the operation speed is limited. Therefore, decimation filters are generally composed of multi-stage filters as shown in Figure 6.35. In this example, the downsampling ratio is 1/8 in the first stage, 1/2 in the next stage, and 1/4 in the last stage, resulting in a total decimation ratio of 64. The first stage needs to operate at the oversampling frequency and has the anti-aliasing characteristics. The first stage, therefore, dominates the performance of the entire decimation filter. In general, decimation is performed with a comb filter having a small number of taps in the first stage, and an FIR filter with high accuracy is used for subsequent stages where the bit rate decreases. In the following description, the comb filter used in the first stage is described.

A simple example of the first stage in a decimation filter is a moving average $w(n)$ of successive N values. Assuming that the output of the $\Delta\Sigma$

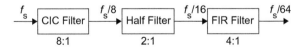

Figure 6.35 Typical decimation filter configuration.

modulator is $v(n)$, $w(n)$ can be written as

$$w(n) = \frac{1}{N}\sum_{i=0}^{N-1}v(n-i). \tag{6.45}$$

Taking the z-transform yields

$$W(z) = \frac{1}{N}(1 + z^{-1} + z^{-2} + \cdots + z^{-(n-1)})V(z). \tag{6.46}$$

Therefore, the transfer function is

$$H_1(z) = \frac{1}{N}\frac{1 - z^{-N}}{1 - z^{-1}}. \tag{6.47}$$

If it is expressed in the frequency domain, its absolute value is

$$|H_1(e^{j2\pi fT_s})| = \left|\frac{1}{N}\frac{1 - e^{-j2\pi NfT_s}}{1 - e^{-j2\pi fT_s}}\right| = \frac{\text{sinc}(NfT_s)}{\text{sinc}(f)}. \tag{6.48}$$

This is the sinc[1] filter characteristics. Figure 6.36 shows the characteristics for $N = 8$. By sampling every eight outputs from the filter, the downsampling with a decimation ratio of $1/8$ can be achieved. As shown by Figure 6.36, since the zeros of the filter are located at frequencies of integer multiples of $f_s/8$, it is possible to prevent high-frequency components from being

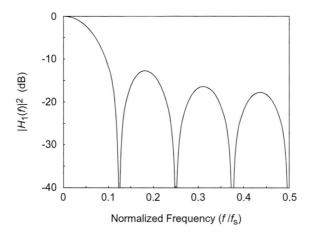

Figure 6.36   Characteristics of sinc[1] filter with $N = 8$.

folded back onto the low-frequency signal band. In general, a filter consisting of cascade-connected k sinc^1 filters is called a sinc^k filter, whose transfer function is

$$H_k(z) = \left(\frac{1}{N} \frac{1-z^{-N}}{1-z^{-1}} \right)^k.$$ (6.49)

Next, consider the order of the filter required when using it as a decimation filter. When using the sinc^1 filter, the noise transmitted through the filter can be written as

$$Q_1(z) = H_1(z)NTF(z)E(z) = \frac{1}{N} \frac{1-z^{-N}}{1-z^{-1}} (1-z^{-1})E(z)$$

$$= \frac{1}{N}(1-z^{-N})E(z),$$ (6.50)

so that the noise power can be obtained as

$$q_1(n) = \frac{1}{N}(e(n) - e(n-N)) \;\rightarrow\; \overline{q_1^2} = \frac{2\overline{e^2}}{N}.$$ (6.51)

Also, when using the sinc^2 filter, the noise passing through the filter can be written as

$$Q_2(z) = H_2(z)NTF(z)E(z) = \left[\frac{1}{N} \frac{1-z^{-N}}{1-z^{-1}} \right]^2 (1-z^{-1})E(z)$$

$$= \frac{1}{N}H_1(z)(1-z^{-N})E(z),$$ (6.52)

so that the noise power can be obtained as

$$q_2(n) = \frac{1}{N^2} \sum_{i=0}^{N-1} [e(n-i) - e(n-N-i)]$$

$$\rightarrow\; \overline{q_2^2} = \frac{2N\overline{e^2}}{N^4} = \frac{2\overline{e^2}}{N^3}.$$ (6.53)

The signal-band noise power of the first-order $\Delta\Sigma$ modulator is $\frac{\pi^2 e_{rms}^2}{3(OSR)^3}$ (Equation (6.12)). Therefore, if the sinc^2 filter is used, the quantization noise

(a)

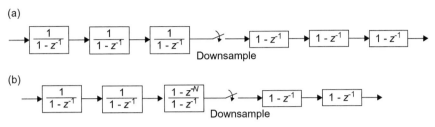

(b)

Figure 6.37   Block Diagrams of the CIC Filter.

power of the signal band and the noise power due to the decimation are of the same order. Here, $N = \mathrm{OSR}$ is used. In general, if L is the order of the $\Delta\Sigma$ modulator, the noise intensity due to decimation will be the same as the quantization noise intensity of the $\Delta\Sigma$ modulator of $k = L + 1$. In other words, the required filter order k is $L + 1$.

A cascaded integrator-comb (CIC) filter described below is often used in circuit implementations instead of a cascaded connection of sinc filters. Block diagram of the CIC filter [136] is shown in Figure 6.37. The case of a sinc³ filter is shown as an example. The denominator of the sinc filter means integration, while the numerator means the subtraction of the delayed signal from the original one. Thus, the CIC filter consists of the first stage performing integration, the second stage performing subtraction, and a down sampler placed between them. The transfer function is

$$H(z) = \left[\frac{1 - z^{-M}}{M(1 - z^{-1})} \right]^3$$

$$= \left(\frac{1}{1 - z^{-1}} \right) \left(\frac{1}{1 - z^{-1}} \right) \left(\frac{1}{1 - z^{-1}} \right)$$

$$\times \left(\frac{1 - z^{-M}}{M} \right) \left(\frac{1 - z^{-M}}{M} \right) \left(\frac{1 - z^{-M}}{M} \right). \qquad (6.54)$$

The generalized transfer function can be expressed as

$$((1 - z^{-M})/(1 - z^{-1}))^k = (1 + z^{-1} + z^{-2} + \cdots + z^{-(M-1)})^k. \qquad (6.55)$$

Here, k is the order of the filter, and M is the decimation ratio. The left side of this equation represents the combination of the integrator and the comb filter, while the right side corresponds to the FIR filter with $k(M - 1)$ taps.

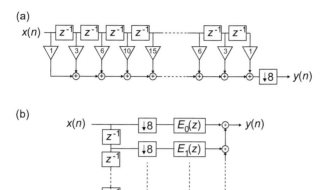

Figure 6.38 (a) Conventional FIR filter and (b) polyphase FIR filter.

The bit width N required for the integrator can be expressed as [136]

$$N = m + k \log_2 M, \tag{6.56}$$

where m is the input bit width, usually 1.

Finally, the polyphase decimation filter [137] proposed as a configuration suitable for low power consumption is explained. Here, there is no need for circuits to operate at the oversampling rate except for the delay elements. Assuming $k = 3$ and $M = 8$, expanding the righthand side of Equation (6.55) yields

$$H(z) = 1 + 3z^{-1} + 6z^{-2} + 10z^{-3} + 15z^{-4} + 21z^{-5} + \cdots$$
$$+ 15z^{-17} + 10z^{-18} + 6z^{-19} + 3z^{-20} + z^{-21}. \tag{6.57}$$

When rearranging the terms on the right side, it can be expressed as

$$H(z) = (1 + 42z^{-8} + 21z^{-16}) + z^{-1}(3 + 46z^{-8} + 15z^{-16}) + \cdots$$
$$+ z^{-6}(28 + 36z^{-8}) + z^{-7}(36 + 28z^{-8}). \tag{6.58}$$

An example of a polyphase FIR filter based on this expression is shown in Figure 6.38. In the polyphase circuit, only the delay element needs to operate at a high oversampling rate, and the remaining circuits operate at 1/8 of the rate. Therefore, low power consumption can be expected. A possible further speeding-up has been proposed by operating it in a time-interleave manner instead of the delay element [138].

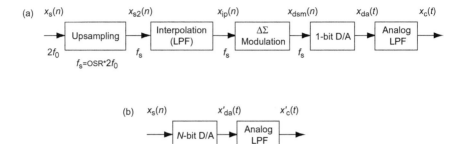

Figure 6.39 (a) $\Delta\Sigma$ type DAC and (b) conventional DAC.

6.8 Oversampling D/A Converter

The $\Delta\Sigma$ modulator is used for not only A/D converters but also D/A converters[7]. A block diagram of a $\Delta\Sigma$-type D/A converter is shown in Figure 6.39 as compared with a conventional one. Notice that unlike the $\Delta\Sigma$-type A/D converter described above, the $\Delta\Sigma$ modulator here is composed of digital circuits. The signal path in Figure 6.39(a) seems more complicated than that in Figure 6.39(b). However, regarding the analog parts, it is not the case. In Figure 6.39(a), the analog signal is processed only on the last two blocks, where the analog signal is represented by the 1-bit signal that is inherently linear. On the other hand, the conventional type in Figure 6.39(b) relies on the N-bit signal, and the linearity of the N-bit D/A converter is a significant issue.

Let us explain the process of D/A conversion by using Figure 6.39(a) and Figure 6.40. First, by upsampling the input digital value $x_s(n)$, $x_{s2}(n)$ is obtained. Next, interpolation using digital circuitry determines the upsampled values, resulting in $x_{ip}(n)$. As shown in the spectrum of Figure 6.40, this plays the role of an analog lowpass filter (LPF) in the conventional configuration. Processing in the digital domain guarantees high linearity. The digital data is then applied to the digital $\Delta\Sigma$ modulator, the output from which is the 1-bit signal $x_{dsm}(n)$. Although it is performed in the digital domain, the input/output characteristics are the same as the analog $\Delta\Sigma$ modulator described in the previous sections, which is shown in Figure 6.40. This 1-bit

---

[7]The $\Delta\Sigma$ modulator digitally expresses the analog signal, and it is often used in scenes where it is necessary to digitally process analog information, such as a PLL (phase locked loop) circuit and an artificial neural network. It is expected that such applications will further expand in the future.

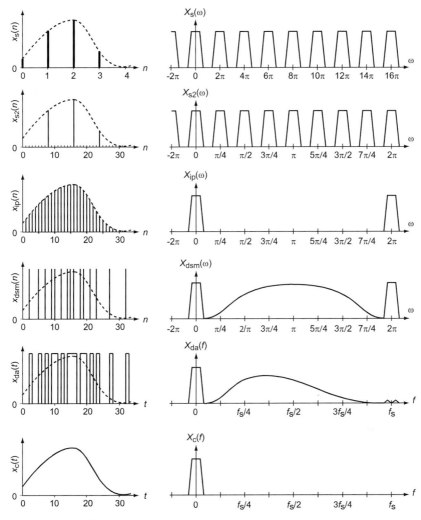

Figure 6.40 Waveforms in $\Delta\Sigma$ type D/A converter.

signal is then applied to the 1-bit D/A converter, followed by the analog LPF, resulting in the final analog output $x_c(t)$. Since the sampled values are held until the next sampling, $X_{\mathrm{dsm}}(\omega)$ is multiplied by the sinc characteristic as shown by $X_{\mathrm{da}}(f)$. The 1-bit signal is inherently linear, and it is possible to avoid the problem of nonlinearity associated with multi-bit D/A converters used in the conventional configuration.

In the conventional multi-bit D/A converter shown in Figure 6.39(b), a steep cutoff characteristic is required for the analog LPF to block high-frequency signal components existing adjacent to the signal band as indicated by $X_s(\omega)$ in Figure 6.40. On the other hand, in the $\Delta\Sigma$ D/A converter, the requirement for steepness is considerably relaxed because the targeted signal is $X_{da}(f)$ shown in Figure 6.40. A digital interpolation circuit realizes the steep cutoff characteristic required in the conventional D/A converter. It can be said that the stringent requirements for analog circuits are satisfied in the digital domain. Note, however, that it must operate at an oversampling frequency.

COLUMN: $\Delta\Sigma$ or $\Sigma\Delta$?

In addition to the $\Delta\Sigma$ modulator, you might sometimes find the term $\Sigma\Delta$ modulator in the literature. Both are the same in contents, and the only difference is naming. "Which should be used" was sometimes discussed in technical meetings in the past. The latter is attributed to the terminology. For example, consider the RMS (root mean square) method. This means to square sampled values, first, and then to calculate a mean value, and finally to take its square root. In other words, the order in the calculation procedure is opposite to the naming. In the case of the modulator, the integration (Σ) is taken after the difference (Δ). Thus, the term $\Sigma\Delta$ follows the rule. Also, since it is thought of a variation of the Δ modulator, a $\Sigma\Delta$ modulator is appropriate, meaning a Δ modulator with Σ function.

On the other hand, the reason it is called a $\Delta\Sigma$ modulator is, among other things, that the original proposers named it [139]. It is said that it is because they wanted to explicitly show that it is not a simple variation of the Δ modulator, but the invention that sets it apart from the prior art. In the academic field, the original proposal should be respected as much as possible, so in this book, we will adopt the "$\Delta\Sigma$" modulator. In recent years, the name of $\Delta\Sigma$ modulator seems to be adopted in many cases.

7

Trends

Advanced CMOS technology has dramatically improved the digital LSI performance. However, analog circuits are not always benefited as well from the scaled-down technology as digital circuits. The output impedance of miniaturized MOSFETs is reduced, making it more challenging to design high-gain opamps. Also, as the supply voltage decreases, the analog signal must be represented with a reduced voltage swing. While reflecting these digital and analog trends, A/D-converter designs have been changing with many innovative ideas from the viewpoints of circuits and architectures, some of which are introduced in this chapter.

Although it is difficult to cover all the topics, some of them are explained as typical examples, to which are worth paying attention. The purpose of this chapter is not to recommend readers to apply them to their targets. Instead, I believe that understanding such typical cases will help solve the problems they are facing, and such efforts are expected to lead to innovative data converters in the future. It is also interesting to note that the basic concepts behind attractive approaches in recent years can find their roots in earlier literature. Therefore, original proposals are cited as much as possible: *i.e.,* visiting the old days to come across new things.

First, the figure of merit (FOM) to estimate A/D converter performance is described. Next, circuit-level approaches to reduce the power consumption of opamp-based A/D converters, such as pipelined and oversampling A/D converters, are discussed. Then, incorporating a SAR A/D converter into pipelined or noise-shaping architecture is explained as examples which are often referred to as hybrid ADCs. Finally, efforts have been presented that aim to further improve performance by using digitally-assisted analog circuits.

7.1 Figure of Merit (FOM)

Many specifications represent A/D converter performance, such as power consumption P, conversion frequency f, supply voltage V, occupied area A, technology node L, and bit resolution B. It has been attempted to build a function by putting them together to express the performance in a single term. Called a figure of merit (FOM), it is written in general as [140–143]

$$F = K \times P^{\alpha_P} \times f^{\alpha_f} \times V^{\alpha_V} \times A^{\alpha_A} \times L^{\alpha_L} \times 2^{\alpha_B B}. \qquad (7.1)$$

K is a normalization constant. Many FOMs can be formed by various combination and interpretation of these variables. In the following, the two most commonly-used FOMs are explained.

If $K = 1$, $f = f_s$, $B = \text{ENOB}$, $V = A = L = 1$, $\alpha_B = \alpha_f = -1$, and $\alpha_P = 1$,

$$\text{FOM}_W = \frac{P}{2^{\text{ENOB}} \cdot f_s} \qquad (7.2)$$

is obtained. This is known as Walden's FOM [144, 145][1]. The reason for taking this form is to associate the FOM with the power consumption, and a small FOM implies a low power consumption. Since the power P is divided by the sampling frequency f_s, the unit of FOM_W is Ws or J (Joule). Instead, J/conv·step is commonly used to mean the energy necessary for converting the analog input with 1-LSB accuracy.

Equation (7.2) means that improving the resolution by 1 bit doubles the energy needed. This will be understood if one remembers that this FOM was first proposed for a flash A/D converter. For example, as shown in Figure 7.1, consider improving the resolution of a 3-bit flash A/D converter by 1 bit. It is sufficient to add another 3-bit flash converter for obtaining the 4th bit. The circuit scale then almost doubles, and the required energy doubles also.

In order to increase the resolution by 1 bit in the SAR A/D converter, adding another comparison cycle seems sufficient. However, at the same time, it is necessary to double the total capacitance of the built-in D/A converter. For example, to obtain a 5-bit resolution, the capacitors of C_0, $2C_0$, $4C_0$, $8C_0$, and $16C_0$ are required. To obtain a 6-bit resolution, $32C_0$ should be added. Recall that there exists a lower limit on the unit capacitance value C_0 to

---

[1]This formula is already described in the paper on a flash A/D converter published in 1980 [146], but here we will call it Walden's FOM according to a convention. The International Technology Roadmap for Semiconductors (ITRS) also adopted this FOM.

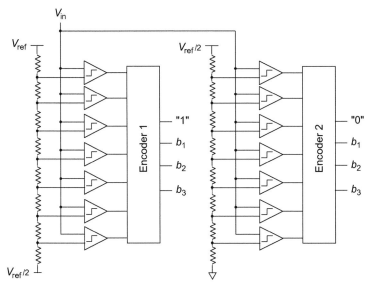

Figure 7.1 Obtaining 4-bit resolution by using two 3-bit flash type A/D converters.

suppress the influence of process variations and thermal noise. Therefore, this leads to a conclusion that the total capacitance doubles, and that the energy required for charging and discharging also doubles.

For additional 1-bit resolution in a 1-bit/stage pipelined A/D converter, one extra pipeline stage is necessary. For each stage, it is necessary to ensure the accuracy corresponding to all subsequent-stage conversions. For example, let us assume that a 5-bit resolution consisting of 5 stages. The conversion in the first stage should be done with 5-bit accuracy. If one attempts to add another stage to make it a 6-bit configuration, the added first-stage conversion should be performed with 6-bit accuracy. Then twice as much capacitance is required in this case as well, and twice as much energy is required for the 1-bit improvement. Therefore, it is reasonable to adopt this FOM not only for the flash but also for other types of A/D converters. It is assumed here that the circuit configuration determines the resolution, and that there is no influence of thermal noise on the resolution.

When the thermal noise dominates the upper limit on the bit resolution, consider another figure of merit expressed as

$$\text{FOM}_{S1} = \frac{2^{2\text{ENOB}} \cdot \text{BW}}{P}. \tag{7.3}$$

This is derived from Equation (7.1) with $K = 1$, $f = $ BW, $B = $ ENOB, $\alpha_B = 2$, $\alpha_f = 1$, $\alpha_P = -1$. Here, BW is the signal bandwidth. The essential difference from Equation (7.2) is to use $2^{2\text{ENOB}}$ rather than 2^{ENOB}. 2^{ENOB} is the ratio of the full-scale signal amplitude to the noise signal amplitude, whereas the squared ratio $2^{2\text{ENOB}}$ is interpreted as the power ratio. In other words, $2^{2\text{ENOB}}$ represents the dynamic range (DR). Therefore Equation (7.3) can be written as

$$\text{FOM}_{\text{S2}} = \frac{\text{DR} \cdot \text{BW}}{P}. \tag{7.4}$$

Taking the logarithm of both sides yields

$$\text{FOM}_{\text{S}} \text{ (in dB)} = \text{DR (in dB)} + 10 \log_{10} \frac{\text{BW}}{P}. \tag{7.5}$$

This is the FOM known as Schreier's FOM [25][2]. It seems strange that the numerator and denominator in the last term logarithm of Equation (7.5) have different dimensions. Nevertheless, the numerical values expressed in units of Hz and W are used as BW and P, respectively, and FOM_{S} in dB is usually calculated. Replacing DR with SNDR gives the FOM taking the distortion into account as [148]

$$\text{FOM}_{\text{S,SNDR}} \text{ (in dB)} = \text{SNDR (in dB)} + 10 \log_{10} \frac{f_\text{s}/2}{P}. \tag{7.6}$$

Equation (7.3) implies that four times the energy is needed to improve the resolution by 1 bit. This suggests that if four identical A/D converters are used simultaneously to sample the input to obtain the four outputs, and if the four outputs are averaged, then the resolution can be improved by 1 bit. Alternatively, if one A/D converter may be operated at four-times higher the sampling frequency while the analog input is unchanged, then the same improvement can be achieved. Recall that, in general, for measurements with random noise, noise can be reduced to $1/\sqrt{N}$ compared to one measurement by performing N measurements. This agrees with the description mentioned above, meaning that this expression is applicable when thermal noise determines the upper limit on resolution.

In order to determine whether it is appropriate to use FOM_{W} or FOM_{S}, it is necessary to consider the data reported in the literature. According to the data reported for the actual A/D converter [149], FOM_{W} reflected the overall

---

[2]An equivalent equation was already presented in the paper on a $\Delta\Sigma$ A/D converter [147].

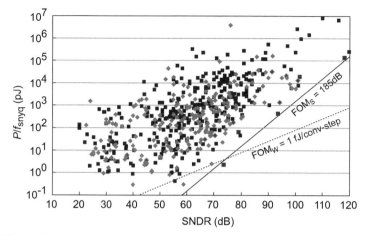

Figure 7.2 Energy necessary for A/D conversion as a function of SNDR [5].

trend until around 2004. Since around 2014, as the power consumption of A/D converters has progressed, it became clear that A/D converters with more than 10 bits of resolution follow FOM_S as is shown in Figure 7.2.

FOM_S depends on the sampling frequency. As shown in Figure 7.3, FOM_S is almost constant in the frequency region below 10–100 MHz, whereas in the higher-frequency region, FOM_S decreases with a slope of approximately 10 dB/dec. The performance in this high-frequency region has been improved in the last ten years. This is partly because of the improved high-frequency characteristics of the MOSFET due to miniaturization, and it is also because of the many valuable ideas introduced in circuit designs. The constant FOM_S in the low-frequency region probably is considered to reflect a theoretical limit determined by Equation (3.31), which is independent of the Nyquist rate $(1/f_{\text{snyq}})$.

Although the FOM is useful as a benchmark to compare the performance of various A/D converters, it is advised to use it in combination with other performance specifications. When showing the performance of the A/D converter, it is necessary to disclose not only the FOM but other useful specifications, such as power consumption, conversion frequency, power supply voltage, occupied area, technology node, and bit resolution. Particular attention is required when comparing A/D converters with different architectures or with significantly different performances.

For further information relating to the FOM, please refer to literature [150–152], as well as those mentioned in Section 1.4.

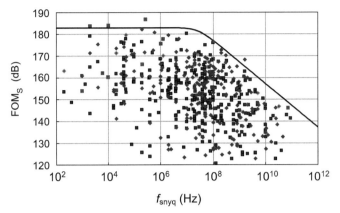

Figure 7.3 FOM$_S$ as a function of sampling frequency [5].

7.2 Low-power Amplifiers

Innovative circuits, as well as scaled-down MOSFETs, have played a critical role to improve the performance of A/D converters. In this section, several circuit examples are explained to reduce the power consumption of opamps, which are used for MDACs in the pipelined A/D converter (Section 5.6) and integrators in the oversampling A/D converter (Chapter 6). This is because the opamp dominates the total power consumption of such opamp-based A/D converters.

The conventional approach for low-power operation of these A/D converters is to reduce the number of opamps. Opamp sharing is introduced, where a single opamp is shared with multiple circuit blocks that require an opamp. For example, the first stage and the second stage in a second-order $\Delta\Sigma$ modulator can share one opamp based on a time-division concept [153]. Also, two adjacent stages in a pipelined A/D converter can share an opamp [154]. However, when the opamp is switched from one circuit block to the other, it takes some time to settle. Also, more switches and wiring increase the complexity of the circuit configuration.

It is attempted to use a comparator instead of an opamp to realize a virtual ground [155]. It has also been investigated to substitute the opamp with a circuit network consisting only of passive elements to perform charge redistribution [156–158]. However, a close look at why opamps are widely used would reveal serious problems to solve in such circuits without opamps; the cost of removing opamps is quite significant. In this section, therefore, amplifiers having a function of suppressing the static current to reduce power consumption is described.

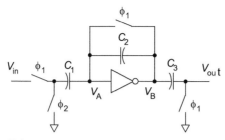

Figure 7.4 Switched-capacitor amplifier using an inverter.

7.2.1 Inverter-based Amplifiers

An example of a switched-capacitor amplifier using inverters [159] is shown in Figure 7.4. In ϕ_1, the circuit is the precharge mode, and $V_A = V_B = V_0$ holds because the input and output of the inverter are connected. Here, V_0 is the logical threshold voltage of the inverter. The charge stored in the right-side electrode of C_1 is $C_1(V_0 - V_{in})$. In ϕ_2, the circuit is the amplification mode. In the transition from ϕ_1 to ϕ_2, no charges flow into or out of the inverter input terminal, because this is a high-impedance node. $V_A = V_0$ holds in the amplification mode because of the feedback via C_2. So, the charge stored in the right-side electrode of C_1 is now C_1V_0. The rest of the charge $-C_1V_{in}$ flows into the left-side electrode of C_2. Since there was no charge stored in C_2 in ϕ_1, the charge in the righthand electrode of C_2 is C_1V_{in}, which leads to the equation

$$V_B = V_0 + \frac{C_1}{C_2}V_{in}. \tag{7.7}$$

It is assumed that V_{out} is connected to the next-stage high-impedance node, so the charge accumulated in C_3 does not change in the transition from ϕ_1 to ϕ_2, which means that the voltage difference across C_3 is V_0. Therefore, in ϕ_2

$$V_{out} = \frac{C_1}{C_2}V_{in} \tag{7.8}$$

holds. V_A is V_0 before and after the charge transfer, which corresponds to the virtual short in conventional opamps. Note that this inverter offset is absorbed by the potential difference across the C_3 so that the input/output characteristics do not depend on V_0.

The inverter through-current flowing from the voltage supply to ground determines the lower limit on the static power consumption. If the inverter operates as a class-C amplifier, the power consumption is much reduced because the MOSFETs go in the weak inversion after the transition. An example of applying an inverter to an integrator of $\Delta\Sigma$ modulator [160, 161] was described in Section 6.3. Also, a two-step hybrid integrator, consisting of a zero-crossing-detector-based integrator and an inverter-based integrator, for IoT sensor applications was proposed [162].

Connecting inverters to form multiple stages increases the gain, which might be desirable for application to an MDAC and integrator. This idea leads to a ring amplifier (RAMP) [104] shown in Figure 7.5. Essentially, this is a three-stage inverter amplifier. However, simply connecting three inverters and applying feedback from the output to the input result in a ring oscillator, so a trick is necessary to suppress the oscillation. As shown in this figure, the second stage consists of two inverters, and an offset voltage of V_{OS} is applied at the inputs of these inverters. In addition, the pMOSFET and the nMOSFET in the third stage are driven separately by signals V_{BP} and V_{BN}. Therefore, as will be described in detail below, after settling, the output impedance of the third stage MOSFET increases because of its weak inversion-region operation, which forms a dominant pole in the low-frequency region, and the circuit is well stabilized.

In Figure 7.5(a), ϕ_{rst} is a switch to reset the circuit, and the offset voltage V_{OS} charges C_2 and C_3. The difference between the logic threshold of the first stage inverter and the input common mode V_{CMX} is canceled by C_1. When feedback is applied to this amplifier as shown in Figure 7.6, V_{in} remains in the vicinity of V_{CMX}. Then, the outputs of the second stage V_{BP} and V_{BN} changes with V_{in} as shown in Figure 7.5(b). Where $V_{OS(IN)}$ is the input-referred offset voltage. In the region of $V_{CMX} \pm V_{OS(IN)}$ (dead zone), both the pMOSFET and nMOSFET in the third stage are in the weak-inversion region so that the output impedance can be high.

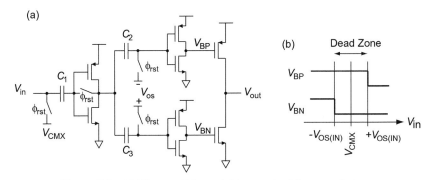

Figure 7.5   (a) Ring amplifier and (b) outputs of the second stage.

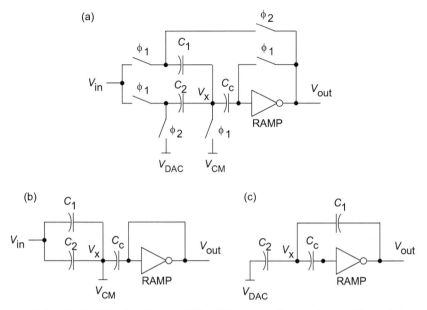

Figure 7.6 (a) MDAC using ring amplifier, (b) its sampling mode and (c) amplification mode.

An MDAC example using a ring amplifier [163] is shown in Figure 7.6. Since there is no charge/discharge of C_c in the transition from the sampling mode to the amplification mode, $V_x = V_{CM}$ holds in the amplification mode. This functions as a virtual ground, and

$$(C_1 + C_2)(V_{CM} - V_{in}) = C_1(V_{CM} - V_{out}) + C_2(V_{CM} - V_{DAC})$$

is obtained. If $C_1 = C_2$, it becomes

$$V_{out} = 2V_{in} - V_{DAC}, \tag{7.9}$$

which shows that this circuit works as an MDAC.

The ring amplifier has attracted attention as a low power consumption amplifier replacing standard opamps. Self-biased ones are applied to pipelined A/D converters [163] [164]. It is also adopted in a pipelined SAR A/D converter [165] and a time-interleaved A/D converter [166].

7.2.2 Dynamic Amplifiers

The inverter-based amplifier described above realized low-power operation by reducing the static current by driving MOSFETs in the weak-inversion region when the charge redistribution is completed. An amplifier that takes this idea a step further is the dynamic amplifier described below. Using MOSFETs as virtually ideal switches the dynamic amplifier further reduces the static current and power consumption.

A prototype of the dynamic amplifier [167] is shown in Figure 7.7. This is a two-stage opamp, which operates in two phases: ϕ_1 and ϕ_2. In ϕ_1, C_0 is discharged. Since ϕ_2 is open, no current flows through the amplifier. By opening ϕ_1 and closing ϕ_2, the circuit goes into the amplification mode. This continues until C_0 is charged such that the voltage across C_0 is V_{DD}. Meanwhile, current flows through the first stage differential pair and the second stage by the current mirror, and the input voltage is amplified. If C_0 is fully charged, the current flowing through this amplifier is 0, and the output terminal will be floating. A switched opamp [168] is also known as an amplifier with a similar configuration.

An example of a dynamic amplifier used in an A/D converter is shown in Figure 7.8. In the circuit shown in Figure 7.8(a), ϕ_1 and ϕ_2 mean the precharge mode and the amplification mode, respectively. In the amplification mode, the output attenuates with time as shown in Figure 7.8(b). This type of dynamic amplifier is used as residue amplifiers in the pipelined A/D converter with 4-channel time-interleave (TI) scheme [169] and the pipelined SAR A/D converter with 2-channel TI scheme [103]. Also reported is a dynamic amplifier [170] that has a common-mode voltage detector (CMD) and can hold the attenuating output at a constant value. An example is shown in Figure 7.8(c), and the outputs change as shown in Figure 7.8(d). When the output common mode voltage reaches a predetermined value, the differential

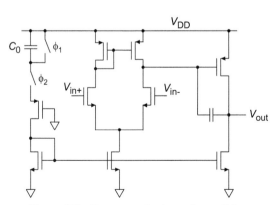

Figure 7.7   Prototype of a dynamic amplifier.

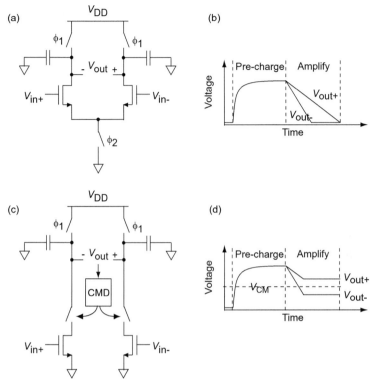

Figure 7.8 (a) Circuit schematic of a conventional dynamic amplifier, (c) that of a dynamic amplifier with common-mode detection, and (b) and (d) output waveforms of (a) and (d), respectively.

output can be kept constant by cutting off the current. This dynamic amplifier is also used as residue amplifiers in the pipelined A/D converter [171], and excellent low-power characteristics are demonstrated.

Dynamic amplifiers using inverters have also been proposed. The amplifier automatically turns off at the end of charge redistribution without using any switches and holds the output. An integrator using such a dynamic amplifier [172] is shown in Figure 7.9. First, ϕ_1 is closed to sample the input V_{in} with C_s. Next, ϕ_2, ϕ_3 and ϕ_5 are closed, and the circuit is in the precharge mode. Then, ϕ_3 is opened and ϕ_4 is closed so that M_1 turns on, and V_{out} decreases as C_o is discharged. At the same time, the gate-to-source voltage V_{GS1} of M_1 also decreases until V_{GS1} falls to the threshold voltage V_t. At that time, M_1 turns off and the charge redistribution is completed.

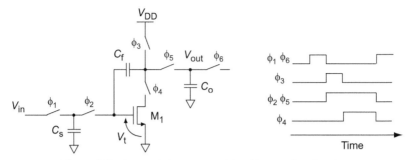

Figure 7.9 Dynamic integrator with self-saturation function.

During the charge redistribution, a part of the charge accumulated in C_s is transferred to C_f. When M_1 turns off, the charge remaining in C_s is constant because the voltage across C_s always settles to V_t. Also, the gate of M_1 is a high impedance node, and there is no charge flowing into or out of this gate. The charge transferred to C_f is $C_i(V_{in} - V_t)$, which is added to the charge that was previously stored in C_f. Therefore, the charge accumulated in C_f determines V_{out}, which indicates that this circuit operates as an integrator. Since there is no path through which the current steadily flows from the power supply to ground, no static power is consumed in this circuit.

A dynamic common source (DCS) amplifier, a circuit that shares a similar operation concept, was proposed [173]. Figure 7.10 shows two integrators using the DCS amplifier: (a) One is using a DCS amplifier, and (b) the other is with a feedback path from a D/A converter so that it can be incorporated in a $\Delta\Sigma$ modulator.

In Figure 7.10(a), ϕ_1 is the sampling phase and the charge $C_s(V_{CM} - V_{in})$ is stored on the right-side electrode of C_s. Diode-connected M_p is initially in the ON state, but eventually, it turns off when the charge equal to $C_{lsp}(V_{DD} - V_{tp} - V_{CM})$ accumulates in C_{lsp}. Here, V_{CM} is the common mode voltage and is assumed to be $V_{DD}/2$. ϕ_2 represents the integration phase. The voltage of the left-side electrode of C_s decreases from $V_{in}(> 0)$ to 0. Accordingly, V_x decreases. Since the charge in C_{lsp} is preserved, the gate voltage of M_p also goes down, M_p turns on, and the drain current flows. Then V_x rises to V_{CM} via C_i, and M_p turns off again, and the charge in C_s is transferred to C_i, and an integration operation can be realized.

The circuit shown in Figure 7.10(b) is an integrator with a feedback path from a D/A converter so that it can be applied to the $\Delta\Sigma$ modulator. When the feedback from the D/A converter is "0", the nMOSFET part (lower half of the circuit) is disconnected, and the circuit becomes the same as in Figure 7.10(a).

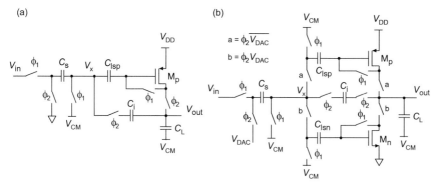

Figure 7.10 (a) Integrator based on a dynamic common-source amplifier and (b) modified one including a feedback signal from a D/A converter for the use in $\Delta\Sigma$ modulators.

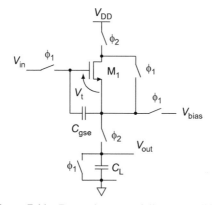

Figure 7.11 Dynamic source follower amplifier.

When the feedback from the D/A converter is "1" (V_{DD}), a drain current for allowing charge redistribution is supplied from the nMOSFET. This circuit is more suitable for low-power operation compared with that shown in Figure 7.9 because it is not necessary to charge C_o in Figure 7.9.

The last topic in this section is the dynamic source follower (DSF) [174] shown in Figure 7.11. ϕ_1 is the sampling phase and the input voltage is sampled with C_{gse}. ϕ_2 is the amplification phase, and the drain current flows until the gate-to-source voltage of M_1 reaches the threshold voltage V_t by charging C_L. At this time, the output voltage V_{out} is obtained based on this charge redistribution. The constant threshold voltage V_t corresponds to the virtual ground of conventional opamps. An MDAC [174, 175] and an integrator for a $\Delta\Sigma$ modulator [176] using DSF circuits have been reported. For details, please refer to the cited documents.

7.3 Hybrid A/D Converters

A/D converters based on various architectures are described in Chapters 5 and 6. Although each architecture was developed almost individually, configurations combining them have also been considered [177–179]. In recent years, progress has been made along the direction with the cutting-edge CMOS technology. This section describes hybrid A/D converters based on successive approximation (SAR) combined with pipelined or noise-shaping architecture. The purpose of these is to realize high resolution and high speed while maintaining low-power consumption of SAR A/D converters. In SAR A/D converters, quantization errors or residues remain as the charge in capacitors of the built-in D/A converter. Therefore, the SAR architecture works well with MDACs for pipelined A/D converters and integrators for the oversampling $\Delta\Sigma$ A/D converters. Circuits can be implemented by modifying the wiring and switches as described below.

7.3.1 Pipelined SAR A/D Converters

As described in Section 5.6, the pipelined A/D converter uses sub-A/D converters in a pipelined manner for satisfying both high-speed and high-resolution operation simultaneously. Connecting K sub-A/D converters with M-bit/stage resolution yields a resolution of $N(=\ M \times K)$ bits. Conventionally, a small value of M is preferred, and a configuration in which 1.5-bits/stage sub-A/D converters are connected to form a ten-stage pipeline, for example, has been widely adopted as mentioned in the previous section.

Pipelined successive-approximation (SAR) A/D converters [180, 181] is based on a different design concept: The resolution of each stage increases, while the number of stages decreases. The dramatic improvement in the operating speed of SAR A/D converters by using the scaled-down CMOS technology has made it practical to replace low-resolution flash A/D converters with high-resolution SAR A/D converters. The purpose is to maintain its excellent low-power characteristics and at the same time to increase the bit resolution. Considering that power-hungry opamps are necessary for residue amplification, it is preferable that the number of stages is small from the viewpoint of low-power consumption operation. Two- or three-stage pipelined SAR A/D converters have been typically designed.

A multi-bit output from the first stage requires a substantial gain for the residue amplifier. If the output of the first stage is M bits, a gain of 2^M is required. At the same time, it can relax the MDAC requirements. For example, if the initial stage is set to 2 bits in order to obtain a 12-bit resolution, the accuracy of the remaining 10 bits is required for the MDAC. On the other hand, if, for example, the first stage has a 6-bit configuration, the accuracy required for the MDAC is 6 bits. Therefore, settling conditions can be significantly relaxed.

Even if the resolution is the same as that of a conventional SAR A/D converters, there is also a merit that the total capacitance can be significantly reduced by adopting a two-stage pipelined configuration. In order to reduce the mismatch and suppress the influence of thermal noise, the unit capacitance value has a lower limit. Therefore, for obtaining high resolution, the total capacitance of the D/A converter has to be increased, and so does the input capacitance. If the pipelined architecture is used, it is possible to reduce input capacitance as well as total capacitance. Conventionally, the total capacitance required for a resolution of N bits is 2^N times the unit capacitance, but by using a two-stage pipeline with $N/2$ bit resolution, the total capacitance can be reduced to $2 \times 2^{N/2}$ times the unit capacitance. The operation is speeded up by reducing the input capacitance, and it also leads to the low power consumption of the input buffer[3].

A circuit example of 4-bit SAR A/D converter assumed to be used in a pipelined structure [180] is shown in Figure 7.12. This converter operates in three modes: sampling, SAR conversion, and residue amplification. In the sampling mode, the input V_{in} is sampled with the DAC capacitors. Switches b_1 through b_4 are connected to V_{in}, and SW to ground. b_0 is closed so that V_x is the virtual ground. In the SAR conversion mode, a conventional SAR switching operation is performed. While SW is connected to ground, b_0 is opened so that the opamp is used as a preamplifier for the comparator. In the residue amplification mode, C_f is used as a feedback capacitor by turning SW to the output of the opamp[4]. Then, it becomes a switched-capacitor amplifier, and the amplified residue is obtained as $V_{residue}$, which is the input to the next stage.

Further speeding up is possible by time-interleaving the SAR A/D converter. A sampling frequency as high as 1.35 GS/s was achieved by using six channels of 2-stage pipelined SAR A/D converters [182]. It is essential to reduce the power consumption of opamps for the residue amplification to exploit the low-power performance unique to the SAR A/D converter. For that

---

[3]This is the same situation that the number of comparators can be reduced when the flash type was changed to two steps.

[4]For low-power consumption, the opamp can be separated from the signal path in the SAR conversion mode, and used only in the residual amplification mode [181].

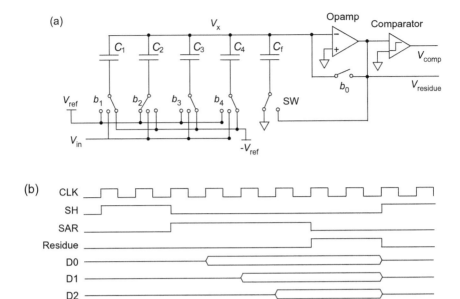

Figure 7.12   (a) Circuit schematic of a pipelined SAR and (b) timing chart.

purpose, the dynamic amplifier [103, 183, 184] and the ring amplifier [165] mentioned in the previous section are adopted. Also, a passive circuit [185] is adapted for residue transfer. There are other encouraging examples reported [186–191], so readers interested should refer to those papers.

7.3.2 Noise-shaping SAR A/D Converters

The first proposal, to the best of the author's knowledge, incorporating the noise shaping into SAR A/D converter is shown in Figure 7.13 [192]. In the conventional SAR operation, when the previous conversion is completed, and the conversion of the next sampled value starts, all charges left in capacitors are discarded. Instead of discarding, sampling and holding the charges enable the noise shaping. The z-transform representation is shown in Figure 7.13(b), which shows that shaping of $(1 + z^{-1})^{-1}$ can be obtained. However, unlike the first-order shaping characteristic $(1 - z^{-1})$ mentioned in Section 6.2, the noise shaping characteristics are not satisfactory, because $z = 1$ (dc input) is not a pole. By adding an integrator, the first-order shaping characteristics were realized, which was demonstrated by circuit fabrication [192].

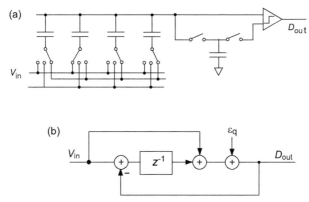

Figure 7.13 (a) Attempt to implement a noise-shaping SAR A/D converter and (b) signal-level block diagram.

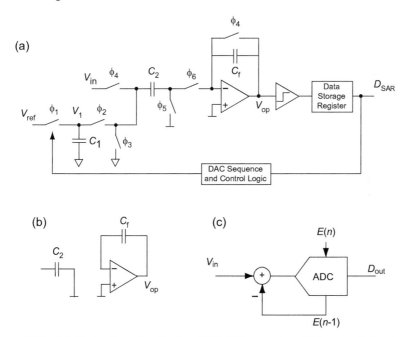

Figure 7.14 (a) First-order noise-shaping SAR A/D converter, (b) its essential part, and (c) simplified block diagram.

A circuit including an integration function in SAR A/D converter [193] is shown in Figure 7.14. This circuit employs a serial D/A converter consisting of two capacitors with equal capacitance values C_1 and C_2. To operate the circuit as an A/D converter, first, ϕ_4, ϕ_5 is HIGH so that input V_{in} is sampled with C_2, and C_f is discharged. At the same time, ϕ_1 is HIGH to charge C_1 with V_{ref}. Next, the sampled input charge is transferred to C_f by setting ϕ_1, ϕ_4, ϕ_5 to LOW, and ϕ_3, ϕ_6 to HIGH. At this time, the output V_{op} of the opamp is equal to V_{in}.

Then, the circuit is in the SAR-conversion mode and determines the MSB. First, when ϕ_6 is kept HIGH, ϕ_3 is set to LOW, and ϕ_2 is set to HIGH, then V_1 equals to $V_{ref}/2$. At the same time, the charge of $C_2 V_{ref}/2$ is transferred to C_f, the comparator input is $V_{op} = V_{in} - V_{ref}/2$, and the sign of which determines the MSB. If the MSB is 0, $C_2 V_{ref}/2$ must be extracted from C_f. It is achieved by the following procedure: ϕ_6 is left HIGH, ϕ_2 is set to LOW, and ϕ_3 is set to HIGH. Then $V_{op} = V_{in}$ is obtained, which allows the circuit to proceed to the second MSB (MSB-1) determination. If the MSB is 1, the circuit can proceed to the MSB-1 determination without any changes. For the MSB-1 decision, ϕ_6 is first set to LOW, ϕ_3 and ϕ_5 are set to HIGH to discharge C_2. Next, setting ϕ_2 and ϕ_6 to HIGH results in $V_1 = V_{ref}/4$. Then, the electric charge of $C_2 V_{ref}/4$ is transferred to C_f to determine MSB-1. The SAR conversion is continued similarly.

For obtaining noise shaping characteristics, it is essential that the quantization error for the previously sampled value is stored in the feedback capacitor:

$$V_{op} = V_{in}(n-1) - V_{DAC}(n-1) \tag{7.10}$$

holds in Figure 7.14(b). If the next conversion is performed without discharging the feedback capacitor, as shown in Figure 7.14(c), the value subtracted by

$$E(n-1) = V_{DAC}(n-1) - V_{in}(n-1) \tag{7.11}$$

from the next input is converted in the next cycle. Thus,

$$D_{out}(n) = V_{in}(n) + E(n) - E(n-1) \tag{7.12}$$

is satisfied, which means that the shaping characteristics represented as $1 - z^{-1}$ is realized. This is just the first-order shaping characteristic.

As with the pipelined SAR A/D converters, opamps can increase power consumption. Employing a passive circuit [194] has been reported to suppress an increase in power consumption. Another attempt is to obtain a third-order noise shaping configuration [195] by adopting the DEM method. This configuration is equivalent to replacing the quantizer of $\Delta\Sigma$ modulator called error feedback type [23] with SAR A/D converter. In this configuration, since an analog integrator exists in the feedback path and its output is directly subtracted from the input signal, its imperfection causes SNR degradation. For this reason, it was rarely adopted in the conventional 1-bit $\Delta\Sigma$ modulator. However, since the 1-bit quantizer is replaced with a SAR A/D converter, the quantization error can be reduced, which makes it possible that this shortcoming can be overcome.

7.4 Digitally-assisted Calibrations

Historically, laser trimming is known as a tool used for achieving high-precision D/A converters [196]. Since the circuit is individually trimmed while monitoring the resistance value with an on-wafer prober, the accuracy is high, but the productivity is low, and the process is expensive. Also, it cannot cope with time-dependent changes in the operating environment. In recent years, techniques that complement the weaknesses of analog circuits by utilizing digital circuitry attract increasing attention.

In this book, several related topics have been already explained. The dynamic element matching (DEM) to suppress the effect of a current-source mismatch has already been described in Sections 4.5 and 6.5[5]. It was also explained that the Wallace Tree eliminates bubble errors in the flash A/D converter (Section 5.2) and that the redundancy in the SAR (Section 5.4) and pipelined (Section 5.6) A/D converters correct decision errors. These were useful calibration methods to reduce conversion errors by subsequent signal manipulation even if the quantization error is large or there are some conversion errors[6].

Many alternative approaches have been proposed, where the conversion errors are evaluated by utilizing digital circuitry and by feeding back the result to the converter. A typical block diagram for calibration is shown Figure 7.15 [199]. The digital output of the A/D converter is converted to an analog value again by the D/A converter, which is compared with the input to obtain the conversion error. Correction is carried out in the digital and

---

[5]Interested readers should also refer to other articles [51, 197–199].

[6]There are compensation and correction as words having similar meanings. Compensation implies correcting the error for which the cause is known such as the phase compensation of opamps. Correction is often used to change something wrong to something right in a general sense. On the other hand, we call it calibration to reduce the difference between the evaluated value and the true value in some way without concerning the cause.

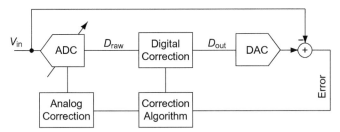

Figure 7.15   Digital calibration model.

analog domains based on a specific correction algorithm. In the following, we introduce such methods of calibrating the data converter with conversion errors due to nonidealities in analog circuits, such as the mismatches between circuit elements.

Calibration methods can be classified into two groups: foreground (FG) calibration and background (BG) calibration. In the FG calibration, the conversion of the actual signal is interrupted for a moment so that the converter can be calibrated by using the test signal. The FG calibration is performed when the power supply is turned on or when a change in the operating environment is detected. The BG calibration is carried out without any interruptions. The BG calibration might seem to be preferable because no conversion interruption is needed. However, it cannot be guaranteed automatically that the convergence and tracking times are predictable regardless of the nature of the input signal. For example, when the signal intensity distribution is not uniform, when a specific periodic wave is included in the signal, or when the input range is exceeded, a problem might occur in the convergence properties. At present, either one is not obviously superior to the other, and both are used in various situations.

7.4.1 Foreground Calibrations

7.4.1.1 SAR A/D converter

One of the most interesting calibrations known for SAR A/D converters [66] is shown in Figure 7.16. If there is no capacitance mismatch in the D/A converter using binary-weighted capacitors, as described in Section 5.4, the value of capacitor C_1 used for the most significant bit (MSB) decision is equal to the sum of other capacitors from C_2 to C_{NB}. In reality, they are not equal to each other due to mismatches. The difference is evaluated by using the calibration D/A converter (CalDAC) shown in the figure and stored in the data register. During the conversion, the data is read out to apply to the

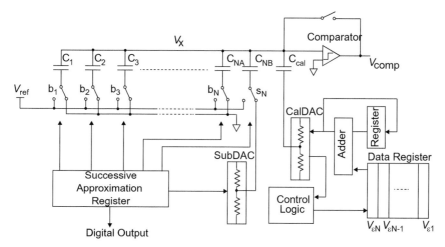

Figure 7.16 Digital calibration for a SAR A/D converter.

CalDAC, and then the analog value generated by the CalDAC corrects the voltage V_x. Decision errors due to capacitance mismatch are thus corrected. The capacitance mismatch occurring in the integrated circuit manufacturing process is about 0.1%, which corresponds to almost 10 bits resolution. It was reported that the resolution could be improved up to 15 bits by this calibration method [66]. In the following, the procedure for mismatch measurement and feedback to the voltage V_x is described for the MSB decision.

First, as shown in Figure 7.17(a), C_1 is connected to ground, and the capacitors from C_2 to C_{NB} are connected to V_{ref}. In the figure, the latter capacitors are collectively represented as C_{2-N}. In this reset mode, the inverting input of the comparator is connected to the output, so that $V_x = 0$ holds, which is similar to the virtual ground of an opamp. Next, as shown in Figure 7.17(b), C_1 is connected to V_{ref} and C_{2-N} is connected to ground. Since the inverting input of the comparator is a high impedance node, the total charge stored in the upper electrodes of C_1 and C_{2-N} does not change in the transition from (a) to (b). Therefore,

$$-C_{2-N}V_{ref} = C_1 \left(V'_x - V_{ref} \right) + C_{2-N}V'_x, \tag{7.13}$$

and from this equation, the voltage V'_x of the inverting input can be found as

$$V'_x = \frac{C_1 - C_{2-N}}{C_1 + C_{2-N}}V_{ref}. \tag{7.14}$$

If there is no mismatch and $C_1 = C_{2-N}$, $V'_x = 0$.

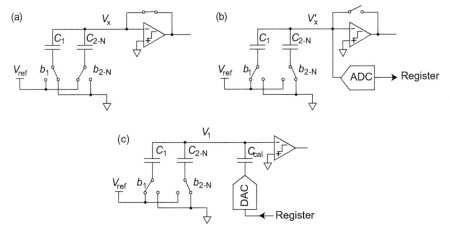

Figure 7.17 Calibration procedure for the MSB (see text).

Let us consider the case where $C_1 \neq C_{2-N}$ due to mismatches. If the variation from C_1's ideal value is ΔC_1, it can be written as

$$C_1 = \frac{C_{\text{total}}}{2} + \Delta C_1 \tag{7.15}$$

$$C_{2-N} = \frac{C_{\text{total}}}{2} - \Delta C_1. \tag{7.16}$$

Here, C_{total} is the sum of all the capacitance values from C_1 to C_{NB}. Substituting these expressions into Equation (7.14) and rearranging the terms result in

$$V'_{\text{x}} = \frac{2\Delta C_1}{C_{\text{total}}} V_{\text{ref}}. \tag{7.17}$$

This value is stored in the data register after D/A conversion.

For the MSB determination, when C_1 is connected to V_{ref} as shown in Figure 7.17(c),

$$V_1 = \frac{C_1 + \Delta C_1}{C_{\text{total}}} V_{\text{ref}} = \frac{1}{2} V_{\text{ref}} + \frac{\Delta C_1}{C_{\text{total}}} V_{\text{ref}} \tag{7.18}$$

is obtained. The second term on the right side is the variation in the threshold voltage due to the capacitance mismatch. Compared with Equation (7.17), it turns out that this is half of the voltage obtained by the operation mentioned above. In the actual MSB determination, the digital value stored in the register in Figure 7.17(b) is read out, the value corresponding to its half is D/A-converted, which is then supplied to the inverting terminal of the comparator via C_{cal}. Thus, the voltage corrected for the mismatch of C_1 is reproduced as V_1.

Correction for MSB-1 decision is performed as follows. This can be done by repeating the above operation for C_2 and all remaining capacitors C_{3-N}, with C_2 connected to ground. However, at this time, it is necessary to consider the influence of the mismatch of C_1. In other words, if the mismatch of C_2 is ΔC_2, then

$$C_2 = \frac{C_{\text{total}}}{4} + \Delta C_2, \qquad (7.19)$$

and the sum of the remaining capacity is written as

$$C_{3-N} = \frac{C_{\text{total}}}{4} - \Delta C_1 - \Delta C_2. \qquad (7.20)$$

By repeating this procedure, information on mismatch can be saved as digital data.

During actual conversion, only when each capacitor is connected to V_{ref}, data is read from the register, and the voltage at the inverting terminal of the comparator is corrected. As shown in Equation (7.20), the capacitance mismatch for the higher bit is handed over to the lower bit, so that it is necessary for the corrections to be summed up using an accumulation register as shown in Figure 7.16.

7.4.1.2 Pipelined A/D converter

An example of FG calibration in a pipelined A/D converter [200] is described next. The circuit diagram of 1.5 bits/step already shown in Figure 5.43 is posted again as Figure 7.18. For an ideal circuit, the relationship between the input and output can be expressed as

$$V_{\text{in}} = \frac{1}{2}V_{\text{out}} + \frac{1}{2}D_{\text{out}}V_{\text{ref}}. \qquad (7.21)$$

In practice, however, it is necessary to consider nonlinearity and gain error because of nonideal factors such as capacitance mismatches, parasitic capacitance, and finite gain. Let us assume that the nonlinearity is represented by the third-order term. Thus the input-output relation can be modified as

$$V_{\text{in,approx}} = \alpha_1 V_{\text{out}} + \alpha_3 V_{\text{out}}^3 + \frac{1}{2}(1 - \epsilon)D_{\text{out}}V_{\text{ref}}, \qquad (7.22)$$

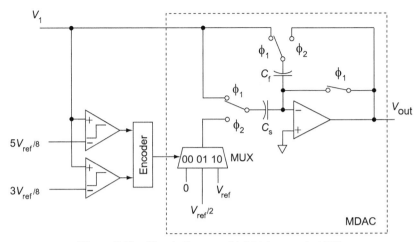

Figure 7.18   Circuit diagram of 1.5-bit/stage sub-ADC.

where ϵ represents a mismatch between the two capacitors, $C_f = C(1 + \epsilon)$, $C_s = C(1 - \epsilon)$. The larger the opamp gain, the closer is the MDAC gain to 2. However, high power consumption is needed to obtain a large gain. Furthermore, it is difficult to increase the output resistance of MOSFETs in the scaled-down CMOS process, so that it is not easy to design an opamp with a large gain. Therefore, not only nonlinearity calibration but also gain calibration is essential, in particular, for the upper bits.

Calibration is carried out from the last stage to the first stage. The method for calibrating the j-th stage is shown in Figure 7.19. Here, the $(j + 1)$-th stage and subsequent stages are assumed to have already been calibrated, and they have ideal characteristics. For calibration, the input analog signal is generated by using a dedicated high precision D/A converter, and the gain w_j for the output $D_{\text{out}.j}$ on the j-th stage is corrected. If there is no gain error, $w_j = (1/2)^j$. Also, for the nonlinearity $f(x)$ occurring at the j-th stage, its inverse characteristic $f^{-1}(x)$ is approximated by the third-order term $\alpha_3 x^3$. The digital output D_{tot} evaluated in this way is compared with the digital input value D_{cal}, and the difference is regarded as an error function. Based on the least-mean-square (LMS) method, w_j and $f^{-1}(x)$ are modified for the error to be minimized.

Figure 7.20 shows the overall configuration consisting of 13 stages of a 1.5-bit resolution and the last stage of a 1-bit resolution. Because the first and second stages have a significant influence on the total resolution, calibration is performed on the residue-amplifier gain error, the D/A-conversion gain

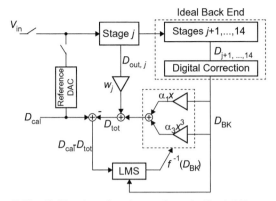

Figure 7.19 Calibration of each stage in a pipelined A/D converter.

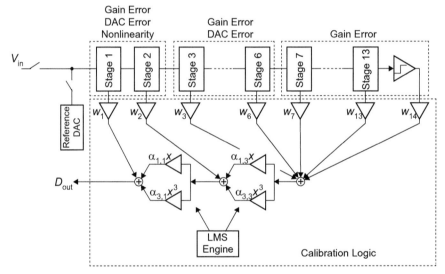

Figure 7.20 Foreground calibration of a pipelined A/D converter.

error, and the nonlinear opamp error. Regarding the next four stages, the residue amplifier gain error and the D/A converter conversion gain error are calibrated. For the remaining seven stages, only the residue gain error is calibrated.

7.4.1.3 Other A/D converters

An example of calibrating the comparator offset in a flash A/D converter is shown in Figure 7.21. The left half of the figure is a comparator composed

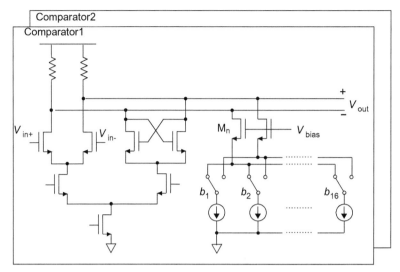

Figure 7.21   Foreground calibration of a flash A/D converter.

of a preamplifier and a latch. The current source group on the right half is a
4-bit current-steering D/A converter. This block is named Comparator1, and
its copy is prepared as Comparator2. First, calibration starts in Comparator1
and applying the same voltage to its positive and negative input terminals
V_{in+} and V_{in-}. Also, all the switches b_i of the D/A converter are turned to the
left terminal as shown in the figure. Since the current from the D/A converter
flows on the left path of the preamplifier, V_- becomes more negative than
V_+ and the differential output V_{out} is positive. Next, consider turning the
switch to the right terminal in order from b_1 to b_2, $\cdots$ b_{16}. Then, the current
flowing in the path on the right side of the preamplifier increases, and V_{out}
decreases and changes its sign from positive to negative after some switching
sequence. By keeping the switch state as it is when the sign V_{out} changes
from positive to negative, the offset of the comparator can be minimized. If
the output does not become negative even if all the switches are turned to
the right, or if the output is negative from the beginning, it is judged that
calibrating Comparator1 is failed, and the same operation is repeated using
the reserved Comparator2. The above calibration method can be considered
as an electrical trimming using the D/A converter. Besides, the calibration
range is expanded by the redundancy of preparing two comparators. By using
this method, it was reported that the linearity and the SNDR of a flash A/D
converter could be improved [201].

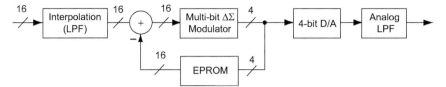

Figure 7.22 Foreground calibration of a $\Delta\Sigma$ D/A converter.

Another example is shown in Figure 7.22 for calibrating a multi-bit $\Delta\Sigma$ D/A converter [202]. In this example, distortion is generated due to the nonlinearity of the 4-bit D/A converter[7]. For calibration, first, using the digitized input of the ramp wave, its nonlinearity is measured off-line in advance. Next, the obtained nonlinearity is stored in the EPROM[8] of the feedback path. If the loop gain is sufficiently large, the input signal and the feedback signal become the same. The nonlinearity is then canceled by the EPROM which is programmed in a such a way that its output is the 16-bit equivalent of any 4-bit input to the D/A converter.

7.4.2 Background Calibrations

A comparator BG calibration for the use in a flash A/D converter [203] is shown in Figure 7.23 with its convergence behavior. V_{in} is a sine wave input with a full scale of V_{FS}, and $V_{\text{R},j}$ is the j-th reference voltage. $q(k)$ is a control signal generated at the k-th clock signal, whose value is randomly selected from ±1. CHP1 is an analog chopper operating with $q(k)$. If $q(k) = 1$, inputs V_{in} and $V_{\text{R},j}$ are transmitted to the subsequent stage as they are, and if $q(k) = -1$, they are exchanged and transmitted to the subsequent stage. Suppose the comparator has an input referred offset voltage V_{OS}. Also, assume that the comparator output D_c is designed to be 1 and 0 if $V_{\text{in}} > V_{\text{R},j}$ and $V_{\text{in}} < V_{\text{R},j}$, respectively.

Background calibration is performed by examining the difference between the distributions of the comparator output D_c when $q(k)$ is 1 and -1. Figure 7.23(b) shows the probability $P(V_{\text{in}} - V_{\text{R},j})(= P(x))$ that the output 1 is obtained as a function of $V_i - V_{\text{R},j}$. If there is a nonzero offset exists, $P(x)$ is independent of $q(k)$. However, if the offset is not 0, the distribution of $P(x)$ depends on $q(k)$. Let P_1 and $P_1 + \Delta P_1$ represent the area where $P(x) = 1$

---

[7]It is described in Section 6.8 to suppress distortion by using a 1-bit D/A converter.
[8]Erasable programmable read-only memory.

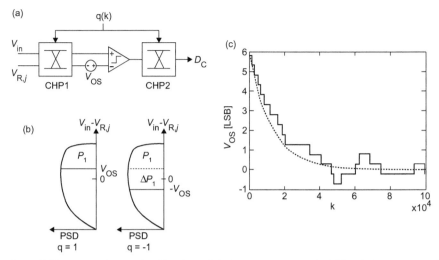

Figure 7.23 (a) Background calibration of flash A/D converter, (b) PSD of the output, and (c) convergence behavior.

for $q = 1$ and $q = -1$, respectively. Then, ΔP_1 determines the sign of the offset voltage. Calibration can be then performed by adjusting the reference voltage of the comparator in the opposite direction to the sign of ΔP_1.

Figure 7.23(c) is a simulation result showing a convergence behavior. The simulation result can be approximated by

$$V_{OS}(k) = V_{OS}(0) \exp\left(-\frac{k}{\tau_c}\right). \tag{7.23}$$

The shorter the time constant τ_c, the better the tracking performance. τ_c depends on the calibration parameters including the threshold for detecting the change in ΔP_1, the variation width of V_{OS} per offset adjustment, and the shape of the probability distribution [203].

Figure 7.24 shows another example of the BG offset calibration [174] of a comparator used for a pipelined A/D converter. This comparator has a pair of differential inputs: One is for the differential analog input and the other is for calibration. Within one sampling period, which includes four phases moving from ϕ_1 to ϕ_4, the comparator makes two decisions at ϕ_1 and ϕ_3. At ϕ_1, the circuit is in the calibration mode and the comparison result $V_{out,comp}$ in the previous phase ϕ_4 is sent to the serial D/A converter consisting of C_{offset} and C_{step}. If C_{step} is connected to and charged by V_{DD} in ϕ_1, V_{offset} will be

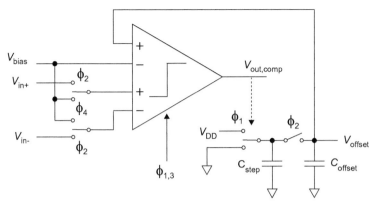

Figure 7.24 Background calibration of comparator offset.

increased by charge redistribution in the next ϕ_2. On the other hand, if C_{step} is connected to GND in ϕ_1, V_{offset} decreases in the next ϕ_2. By changing V_{offset} in this manner, the offset of the comparator is adjusted. The comparison of the input signal, V_{in+} and V_{in-}, connected to the comparator in phase ϕ_2 is done in the compare phase of ϕ_3.

7.4.3 Impact on Designs

The evolution of technology has enabled more powerful digital assistance. However, the careful analog design is still the basis. It should be the role of digital assist to calibrate the imperfections that remain after such efforts.

It is also vital to explore analog designs that are compatible with digital processing. For example, if one relies on traditional analog techniques, a large opamp gain is necessary to obtain exactly a gain of two for the MDAC. As an alternative, it is worthwhile to digitally calibrate the circuit, the gain of which is not precisely two due to a finite opamp gain.

Also, considering that the application fields of A/D converters are getting widespread, it will not be realistic to find powerful calibration means for general purposes. It is inevitable to refine a design that assumes a specific application area and to develop a calibration method optimized for it.

References

[1] W. Kester, "A brief history of data conversion: A tale of nozzles, relays, tubes, transistors, and CMOS," *IEEE Solid-State Circuits Magazine*, vol. 7, no. 3, pp. 16–37, Summer 2015.

[2] D. Robertson, "50 years of analog development at ISSCC," in *2003 IEEE International Solid-State Circuits Conference, 2003. Digest of Technical Papers. ISSCC.*, Feb 2003, pp. 23–24.

[3] ——, "The past, present, and future of data converters and mixed signal ICs: A "universal" model," in *2006 Symposium on VLSI Circuits, 2006. Digest of Technical Papers.*, June 2006, pp. 1–4.

[4] Y. C. Jenq, "Digital spectra of nonuniformly sampled signals: fundamentals and high-speed waveform digitizers," *IEEE Transactions on Instrumentation and Measurement*, vol. 37, no. 2, pp. 245–251, Jun 1988.

[5] B. Murmann, *ADC Performance Survey 1997–2018*. [Online]. Available: http://web.stanford.edu/~murmann/adcsurvey.html.

[6] D. H. Robertson, "Problems and solutions: How applications drive data converters (and how changing data converter technology influences system architecture)," *IEEE Solid-State Circuits Magazine*, vol. 7, no. 3, pp. 47–57, Summer 2015.

[7] R. Gregorian and G. Temes, *Analog MOS integrated circuits for signal processing*, ser. Wiley series on filters. Wiley, 1986. [Online]. Available: https://books.google.co.jp/books?id=GQBTAAAAMAAJ

[8] B. Razavi, *Principles of Data Conversion System Design*. Wiley-IEEE Press, 1995.

[9] R. J. van de Plassche, *CMOS Integrated Analog-to-Digital and Digital-to-Analog Converters, 2nd Ed.* Springer, 2003.

[10] F. Maloberti, *Data Converters*. Springer, 2007. [Online]. Available: https://books.google.co.jp/books?id=Kvo7cjmaEpkC

[11] G. Manganaro, *Advanced Data Converters*. Cambridge, 2013.

[12] M. Pelgrom, *Analog-to-Digital Conversion*. Springer International Publishing, 2016. [Online]. Available: https://books.google.co.jp/books?id=mPQqDQAAQBAJ

[13] M. Rudin, R. O'Day, and R. Jenkins, "System/circuit device considerations in the design and development of a D/A and A/D integrated circuits family," in *1967 IEEE International Solid-State Circuits Conference. Digest of Technical Papers*, vol. X, Feb 1967, pp. 16–17.

[14] B. J. Hosticka, "Performance comparison of analog and digital circuits," *Proceedings of the IEEE*, vol. 73, no. 1, pp. 25–29, Jan 1985.

[15] P. M. Aziz, H. V. Sorensen, and J. van der Spiegel, "An overview of sigma-delta converters," *IEEE Signal Processing Magazine*, vol. 13, no. 1, pp. 61–84, Jan 1996.

[16] S. Rapuano, P. Daponte, E. Balestrieri, L. D. Vito, S. J. Tilden, S. Max, and J. Blair, "ADC parameters and characteristics," *IEEE Instrumentation Measurement Magazine*, vol. 8, no. 5, pp. 44–54, Dec 2005.

[17] K. Bult, "Embedded analog-to-digital converters," in *2009 Proceedings of ESSCIRC*, Sept 2009, pp. 52–64.

[18] L. C. F. Fuiano and P. Charbone, "Data converters: an empirical research on the correlation between scientific literature and patenting activity," in *International Workshop on ADC Modelling, Testing, and Data Converter Analysis and Design and IEEE ADC Forum*, 2011, pp. 1–6.

[19] B. E. Jonsson, "A/D-converter performance evolution," 2013. [Online]. Available: https://pdfs.semanticscholar.org/d754/80e0d260278dd704acbf654440225eb70%c01.pdf

[20] S. Tsukamoto, "Advances in analog-to-digital converters over the last decade," *IEICE Transactions on Fundamentals of Electronics, Communications and Computer Sciences Vol.E100-A No.2*, pp. 524–533, Feb 2017.

[21] W. Kester, Ed., *Data Conversion Handbook*, ser. Analog Devices series. Elsevier, 2005. [Online]. Available: https://books.google.co.jp/books?id=0aeBS6SgtR4C

[22] J. C. Candy and G. C. Temes, *Oversampling Delta-Sigma Data Converters: Theory, Design, and Simulation*. Wiley-IEEE Press, 1992. [Online]. Available: http://ieeexplore.ieee.org/xpl/articleDetails.jsp?arnumber=5312193

[23] S. R. Norsworthy, R. Schreier, and G. C. Temes, *Delta-Sigma Data Converters: Theory, Design, and Simulation.* Wiley-IEEE Press, 1997. [Online]. Available: http://ieeexplore.ieee.org/xpl/article Details.jsp?arnumber=5273727

[24] J. Cherry and W. Snelgrove, *Continuous-Time Delta-Sigma Modulators for High-Speed A/D Conversion: Theory, Practice and Fundamental Performance Limits,* ser. The Springer International Series in Engineering and Computer Science. Springer US, 1999. [Online]. Available: https://books.google.co.jp/books?id= P07fNisLCFoC

[25] R. Schreier and G. C. Temes, *Understanding Delta-Sigma Data Converters.* Wiley-IEEE Press, 2005.

[26] S. Pavan, R. Schreier, and G. Temes, *Understanding Delta-Sigma Data Converters, Second Ed.,* ser. IEEE Press Series on Microelectronic Systems. Wiley, 2017. [Online]. Available: https://books.google.co.jp/books?id=JBauDQAAQBAJ

[27] R. Gregorian, *Introduction to CMOS OP-AMPs and comparators,* ser. A Wiley-Interscience publication. Wiley, 1999. [Online]. Available: https://books.google.co.jp/books?id=uxFTAAAAMAAJ

[28] R. Baker, *CMOS: Circuit Design, Layout, and Simulation,* ser. IEEE Press Series on Microelectronic Systems. Wiley, 2011. [Online]. Available: https://books.google.co.jp/books?id=kxYhNrOKuJQC

[29] T. Carusone, D. Johns, and K. Martin, *Analog Integrated Circuit Design,* ser. Analog Integrated Circuit Design. Wiley, 2012. [Online]. Available: https://books.google.co.jp/books?id=hNvNygAACAAJ

[30] A. Sedra and K. Smith, *Microelectronic Circuits,* ser. Oxford Series in Electrical and Computer Engineering. Oxford University Press, 2014. [Online]. Available: https://books.google.co.jp/books?id=idO-oQEACAAJ

[31] M. Verhelst and A. Bahai, "Where analog meets digital: Analog-to-information conversion and beyond," *IEEE Solid-State Circuits Magazine,* vol. 7, no. 3, pp. 67–80, Summer 2015.

[32] B. Shoop, *Photonic Analog-to-Digital Conversion,* ser. Springer Series in Optical Sciences. Springer Berlin Heidelberg, 2012. [Online]. Available: https://books.google.co.jp/books?id=kRfyCAAAQBAJ

[33] C. Azeredo-Leme, "Clock jitter effects on sampling: A tutorial," *IEEE Circuits and Systems Magazine,* vol. 11, no. 3, pp. 26–37, third quarter 2011.

[34] B. Razavi, "Design of sample-and-hold amplifiers for high-speed low-voltage A/D converters," in *Proceedings of CICC 97 – Custom Integrated Circuits Conference*, May 1997, pp. 59–66.

[35] ——, "The bootstrapped switch [a circuit for all seasons]," *IEEE Solid-State Circuits Magazine*, vol. 7, no. 3, pp. 12–15, Summer 2015.

[36] K. R. Stafford, R. A. Blanchard, and P. R. Gray, "A complete monolithic sample/hold amplifier," *IEEE Journal of Solid-State Circuits*, vol. 9, no. 6, pp. 381–387, Dec 1974.

[37] G. Erdi and P. R. Henneuse, "A precision FET-less sample-and-hold with high charge-to-droop current ratio," *IEEE Journal of Solid-State Circuits*, vol. 13, no. 6, pp. 864–873, Dec 1978.

[38] A. Matsuzawa, M. Kagawa, M. Kanoh, K. Tatehara, T. Yamaoka, and K. Shimizu, "A 10 b 30 MHz two-step parallel BiCMOS ADC with internal S/H," in *1990 37th IEEE International Conference on Solid-State Circuits*, Feb 1990, pp. 162–163.

[39] P. Vorenkamp and J. P. M. Verdaasdonk, "Fully bipolar, 120-Msample/s 10-b track-and-hold circuit," *IEEE Journal of Solid-State Circuits*, vol. 27, no. 7, pp. 988–992, Jul 1992.

[40] A. Moriyama, S. Taniyama, and T. Waho, "A low-distortion switched-source-follower track-and-hold circuit," in *2012 19th IEEE International Conference on Electronics, Circuits, and Systems (ICECS 2012)*, Dec 2012, pp. 105–108.

[41] Y. Lin, H. Chang, and Y. Wang, "Dc-16 GHz GaAs track-and-hold amplifier using sampling rate and linearity enhancement techniques," *Electronics Letters*, vol. 54, no. 2, pp. 83–85, 2018.

[42] M. Dessouky and A. Kaiser, "Very low-voltage digital-audio $\Delta\Sigma$ modulator with 88-dB dynamic range using local switch bootstrapping," *IEEE Journal of Solid-State Circuits*, vol. 36, no. 3, pp. 349–355, Mar 2001.

[43] A. M. Abo and P. R. Gray, "A 1.5-V, 10-bit, 14.3-MS/s CMOS pipeline analog-to-digital converter," *IEEE Journal of Solid-State Circuits*, vol. 34, no. 5, pp. 599–606, May 1999.

[44] C. Svensson, "Towards power centric analog design," *IEEE Circuits and Systems Magazine*, vol. 15, no. 3, pp. 44–51, third quarter 2015.

[45] W. Sansen and C. Svensson, "Comments on the paper "Towards power centric analog design" by Christer Svensson, IEEE Circuits and Systems Magazine, vol. 15, no. 3, pp. 44–51, Sept. 2015. [Express

Letters]," *IEEE Circuits and Systems Magazine*, vol. 16, no. 1, pp. 87–88, First quarter 2016.

[46] A. Yukawa, "A CMOS 8-bit high-speed A/D converter IC," *IEEE Journal of Solid-State Circuits*, vol. 20, no. 3, pp. 775–779, June 1985.

[47] M. van Elzakker, E. van Tuijl, P. Geraedts, D. Schinkel, E. Klumperink, and B. Nauta, "A 1.9 μW 4.4fJ/Conversion-step 10b 1MS/s charge-redistribution ADC," in *2008 IEEE International Solid-State Circuits Conference – Digest of Technical Papers*, Feb 2008, pp. 244–610.

[48] B. Razavi, "The strong ARM latch [a circuit for all seasons]," *IEEE Solid-State Circuits Magazine*, vol. 7, no. 2, pp. 12–17, Spring 2015.

[49] R. E. Suarez, P. R. Gray, and D. A. Hodges, "All-MOS charge-redistribution analog-to-digital conversion techniques. II," *IEEE Journal of Solid-State Circuits*, vol. 10, no. 6, pp. 379–385, Dec 1975.

[50] L. Duncan, B. Dupaix, J. J. McCue, B. Mathieu, M. LaRue, V. J. Patel, M. Teshome, M. Choe, and W. Khalil, "A 10-bit DC-20-GHz multiple-return-to-zero DAC with >48-dB SFDR," *IEEE Journal of Solid-State Circuits*, vol. 52, no. 12, pp. 3262–3275, Dec 2017.

[51] S. M. McDonnell, V. J. Patel, L. Duncan, B. Dupaix, and W. Khalil, "Compensation and calibration techniques for current-steering DACs," *IEEE Circuits and Systems Magazine*, vol. 17, no. 2, pp. 4–26, Second quarter 2017.

[52] B. Razavi, "The current-steering DAC [a circuit for all seasons]," *IEEE Solid-State Circuits Magazine*, vol. 10, no. 1, pp. 11–15, winter 2018.

[53] D. W. J. Groeneveld, H. J. Schouwenaars, H. A. H. Termeer, and C. A. A. Bastiaansen, "A self-calibration technique for monolithic high-resolution D/A converters," *IEEE Journal of Solid-State Circuits*, vol. 24, no. 6, pp. 1517–1522, Dec 1989.

[54] R. J. V. D. Plassche, "Dynamic element matching for high-accuracy monolithic D/A converters," *IEEE Journal of Solid-State Circuits*, vol. 11, no. 6, pp. 795–800, Dec 1976.

[55] J. Briaire, "Error reduction in a digital-to-analog (DAC) converter," Jul. 1, 2008, US Patent 7394414 B2.

[56] D. Tank and J. Hopfield, "Simple 'neural' optimization networks: An A/D converter, signal decision circuit, and a linear programming circuit," *IEEE Trans.Circuits and Systems*, vol. 33, no. 5, pp. 533–541, May 1986.

[57] T. Waho, "A noise-shaping analog-to-digital converter using a $\Delta\Sigma$ modulator feedforward network," *Journal of Applied Logics*, vol. 5, no. 9, pp. 1833–1848, Dec 2018.

[58] M. Choi and A. A. Abidi, "A 6-b 1.3-Gsample/s A/D converter in 0.35-μm CMOS," *IEEE Journal of Solid-State Circuits*, vol. 36, no. 12, pp. 1847–1858, Dec 2001.

[59] I. Dedic, "56Gs/s ADC: Enabling 100GbE," in *2010 Conference on Optical Fiber Communication (OFC/NFOEC), collocated National Fiber Optic Engineers Conference*, March 2010, pp. 1–3.

[60] B. Nauta and A. G. W. Venes, "A 70-MS/s 110-mW 8-b CMOS folding and interpolating A/D converter," *IEEE Journal of Solid-State Circuits*, vol. 30, no. 12, pp. 1302–1308, Dec 1995.

[61] L. Wang, M. LaCroix, and A. C. Carusone, "A 4-GS/s single channel reconfigurable folding flash ADC for wireline applications in 16-nm FinFET," *IEEE Transactions on Circuits and Systems II: Express Briefs*, vol. 64, no. 12, pp. 1367–1371, Dec 2017.

[62] R. van de Grift, I. W. J. M. Rutten, and M. van der Veen, "An 8-bit video ADC incorporating folding and interpolation techniques," *IEEE Journal of Solid-State Circuits*, vol. 22, no. 6, pp. 944–953, Dec 1987.

[63] W. M. Goodall, "Telephony by pulse code modulation," *Bell System Technical Journal*, vol. 26, no. 3, pp. 395–409, 1947. [Online]. Available: http://dx.doi.org/10.1002/j.1538-7305.1947.tb00902.x

[64] S. W. M. Chen and R. W. Brodersen, "A 6-bit 600-MS/s 5.3-mW asynchronous ADC in 0.13-μm CMOS," *IEEE Journal of Solid-State Circuits*, vol. 41, no. 12, pp. 2669–2680, Dec 2006.

[65] J. L. McCreary and P. R. Gray, "All-MOS charge redistribution analog-to-digital conversion techniques. I," *IEEE Journal of Solid-State Circuits*, vol. 10, no. 6, pp. 371–379, Dec 1975.

[66] H. S. Lee, D. A. Hodges, and P. R. Gray, "A self-calibrating 15 bit CMOS A/D converter," *IEEE Journal of Solid-State Circuits*, vol. 19, no. 6, pp. 813–819, Dec 1984.

[67] J. Craninckx and G. van der Plas, "A 65fJ/conversion-step 0-to-50MS/s 0-to-0.7mW 9b charge-sharing SAR ADC in 90nm digital CMOS," in *2007 IEEE International Solid-State Circuits Conference. Digest of Technical Papers*, Feb 2007, pp. 246–600.

[68] B. P. Ginsburg and A. P. Chandrakasan, "An energy-efficient charge recycling approach for a SAR converter with capacitive DAC," in *2005 IEEE International Symposium on Circuits and Systems*, May 2005, pp. 184–187 Vol. 1.

[69] ——, "500-MS/s 5-bit ADC in 65-nm CMOS with split capacitor array DAC," *IEEE Journal of Solid-State Circuits*, vol. 42, no. 4, pp. 739–747, April 2007.

[70] J. Fredenburg and M. P. Flynn, "ADC trends and impact on SAR ADC architecture and analysis," in *2015 IEEE Custom Integrated Circuits Conference (CICC)*, Sept 2015, pp. 1–8.

[71] J.-S. Lee and I.-C. Park, "Capacitor array structure and switch control for energy-efficient SAR analog-to-digital converters," in *2008 IEEE International Symposium on Circuits and Systems*, May 2008, pp. 236–239.

[72] C. Liu, S. Chang, G. Huang, and Y. Lin, "A 0.92mW 10-bit 50-MS/s SAR ADC in 0.13µm CMOS process," in *2009 Symposium on VLSI Circuits*, June 2009, pp. 236–237.

[73] C. C. Liu, S. J. Chang, G. Y. Huang, and Y. Z. Lin, "A 10-bit 50-MS/s SAR ADC with a monotonic capacitor switching procedure," *IEEE Journal of Solid-State Circuits*, vol. 45, no. 4, pp. 731–740, April 2010.

[74] Z. Cao, S. Yan, and Y. Li, "A 32 mW 1.25 GS/s 6b 2b/step SAR ADC in 0.13 µm CMOS," *IEEE Journal of Solid-State Circuits*, vol. 44, no. 3, pp. 862–873, March 2009.

[75] N. Sugiyama, H. Noto, Y. Nishigami, R. Oda, and T. Waho, "A low-power successive approximation analog-to-digital converter based on 2-bit/step comparison," in *2010 40th IEEE International Symposium on Multiple-Valued Logic*, May 2010, pp. 325–330.

[76] Z. Boyacigiller, B. Weir, and P. Bradshaw, "An error-correcting 14b/20µs CMOS A/D converter," in *1981 IEEE International Solid-State Circuits Conference. Digest of Technical Papers*, vol. XXIV, Feb 1981, pp. 62–63.

[77] F. Kuttner, "A 1.2V 10b 20MSample/s non-binary successive approximation ADC in 0.13µm CMOS," in *2002 IEEE International Solid-State Circuits Conference. Digest of Technical Papers (Cat. No.02CH37315)*, vol. 1, Feb 2002, pp. 176–177 vol.1.

[78] T. Ogawa, H. Kobayashi, Y. Takahashi, N. Takai, M. Hotta, H. San, T. Matsuura, A. Abe, K. Yagi, and T. Mori, "SAR ADC algorithm with redundancy and digital error correction," *IEICE Trans. Fundamentals*, vol. E93-A, no. 2, pp. 415–423, Feb 2010.

[79] D. G. Chen, F. Tang, and A. Bermak, "A low-power pilot-DAC based column parallel 8b SAR ADC with forward error correction for CMOS image sensors," *IEEE Transactions on Circuits and Systems I: Regular Papers*, vol. 60, no. 10, pp. 2572–2583, Oct 2013.

[80] H. Fan and F. Maloberti, "High-resolution SAR ADC with enhanced linearity," *IEEE Transactions on Circuits and Systems II: Express Briefs*, vol. 64, no. 10, pp. 1142–1146, Oct 2017.

[81] N. Collins, A. Tamez, L. Jie, J. Pernillo, and M. P. Flynn, "A mismatch-immune 12-bit SAR ADC with completely reconfigurable capacitor DAC," *IEEE Transactions on Circuits and Systems II: Express Briefs*, vol. 65, no. 11, pp. 1589–1593, Nov 2018.

[82] M. Liu, A. H. M. van Roermund, and P. Harpe, "A 7.1-fJ/conversion-step 88-dB SFDR SAR ADC with energy-free "swap to reset"," *IEEE Journal of Solid-State Circuits*, vol. 52, no. 11, pp. 2979–2990, Nov 2017.

[83] W. Guo, Y. Kim, A. H. Tewfik, and N. Sun, "A fully passive compressive sensing SAR ADC for low-power wireless sensors," *IEEE Journal of Solid-State Circuits*, vol. 52, no. 8, pp. 2154–2167, Aug 2017.

[84] T. Waho, "Non-binary successive approximation analog-to-digital converters: A survey," in *2014 IEEE 44th International Symposium on Multiple-Valued Logic*, May 2014, pp. 73–78.

[85] B. Razavi, "A tale of two ADCs: Pipelined versus SAR," *IEEE Solid-State Circuits Magazine*, vol. 7, no. 3, pp. 38–46, Summer 2015.

[86] P. Harpe, "Successive approximation analog-to-digital converters: Improving power efficiency and conversion speed," *IEEE Solid-State Circuits Magazine*, vol. 8, no. 4, pp. 64–73, Fall 2016.

[87] T. Matsuura, "Recent progress on CMOS successive approximation ADCs," *IEEJ Transactions on Electrical and Electronic Engineering*, vol. 11, no. 5, pp. 535–548. [Online]. Available: https://onlinelibrary.wiley.com/doi/abs/10.1002/tee.22290

[88] R. McCharles and D. Hodges, "Charge circuits for analog LSI," *IEEE Transactions on Circuits and Systems*, vol. 25, no. 7, pp. 490–497, July 1978.

[89] S. Masuda, Y. Kitamura, S. Ohya, and M. Kikuchi, "A CMOS pipeline algorithmic A/D converter," *Proc IEEE Custom Integr Circuits Conf*, vol. 1984, pp. 559–562, 1984, a CMOS pipeline algorithmic A/D converter. [Online]. Available: http://jglobal.jst.go.jp/detail.php?JGLOBAL_ID=200902089242583097

[90] S. H. Lewis, "Optimizing the stage resolution in pipelined, multi-stage, analog-to-digital converters for video-rate applications," *IEEE Transactions on Circuits and Systems II: Analog and Digital Signal Processing*, vol. 39, no. 8, pp. 516–523, Aug 1992.

[91] S. H. Lewis and P. R. Gray, "A pipelined 5-Msample/s 9-bit analog-to-digital converter," *IEEE Journal of Solid-State Circuits*, vol. 22, no. 6, pp. 954–961, Dec 1987.

[92] M. Kameyama, M. Nomura, and T. Higuchi, "Modular design of multiple-valued arithmetic VLSI system using signed-digit number system," in *Proceedings of the Twentieth International Symposium on Multiple-Valued Logic*, May 1990, pp. 355–362.

[93] B.-S. Song, M. F. Tompsett, and K. R. Lakshmikumar, "A 12-bit 1-Msample/s capacitor error-averaging pipelined A/D converter," *IEEE Journal of Solid-State Circuits*, vol. 23, no. 6, pp. 1324–1333, Dec 1988.

[94] Y. M. Lin, B. Kim, and P. R. Gray, "A 13-b 2.5-MHz self-calibrated pipelined A/D converter in 3-μm CMOS," *IEEE Journal of Solid-State Circuits*, vol. 26, no. 4, pp. 628–636, Apr 1991.

[95] S. H. Lewis, H. S. Fetterman, G. F. Gross, R. Ramachandran, and T. R. Viswanathan, "A 10-b 20-Msample/s analog-to-digital converter," *IEEE Journal of Solid-State Circuits*, vol. 27, no. 3, pp. 351–358, Mar 1992.

[96] A. N. Karanicolas, H.-S. Lee, and K. L. Barcrania, "A 15-b 1-Msample/s digitally self-calibrated pipeline ADC," *IEEE Journal of Solid-State Circuits*, vol. 28, no. 12, pp. 1207–1215, Dec 1993.

[97] T. B. Cho and P. R. Gray, "A 10 b, 20 Msample/s, 35 mW pipeline A/D converter," *IEEE Journal of Solid-State Circuits*, vol. 30, no. 3, pp. 166–172, Mar 1995.

[98] D. W. Cline and P. R. Gray, "A power optimized 13-b 5 Msamples/s pipelined analog-to-digital converter in 1.2 μm CMOS," *IEEE Journal of Solid-State Circuits*, vol. 31, no. 3, pp. 294–303, Mar 1996.

[99] C. S. G. Conroy, D. W. Cline, and P. R. Gray, "An 8-b 85-MS/s parallel pipeline A/D converter in 1-μm CMOS," *IEEE Journal of Solid-State Circuits*, vol. 28, no. 4, pp. 447–454, Apr 1993.

[100] K. Nakamura, M. Hotta, L. R. Carley, and D. J. Allsot, "An 85 mW, 10 b, 40 Msample/s CMOS parallel-pipelined ADC," *IEEE Journal of Solid-State Circuits*, vol. 30, no. 3, pp. 173–183, Mar 1995.

[101] S. Devarajan, L. Singer, D. Kelly, S. Kosic, T. Pan, J. Silva, J. Brunsilius, D. Rey-Losada, F. Murden, C. Speir, J. Bray, E. Otte, N. Rakuljic, P. Brown, T. Weigandt, Q. Yu, D. Paterson, C. Petersen, and J. Gealow, "A 12b 10GS/s interleaved pipeline ADC in 28nm CMOS technology," in *2017 IEEE International Solid-State Circuits Conference (ISSCC)*, Feb 2017, pp. 288–289.

[102] S. Kawahito, K. Honda, M. Furuta, N. Kawai, and D. Miyazaki, "Low-power design of high-speed A/D converters," *IEICE Transactions on Electronics Vol.E88-C No.4*, pp. 468–478, Apr 2005.

[103] B. Verbruggen, K. Deguchi, B. Malki, and J. Craninckx, "A 70 dB SNDR 200 MS/s 2.3 mW dynamic pipelined SAR ADC in 28nm digital CMOS," in *2014 Symposium on VLSI Circuits Digest of Technical Papers*, June 2014, pp. 1–2.

[104] B. Hershberg, S. Weaver, K. Sobue, S. Takeuchi, K. Hamashita, and U. K. Moon, "Ring amplifiers for switched capacitor circuits," *IEEE Journal of Solid-State Circuits*, vol. 47, no. 12, pp. 2928–2942, Dec 2012.

[105] B. Provost and E. Sanchez-Sinencio, "On-chip ramp generators for mixed-signal BIST and ADC self-test," *IEEE Journal of Solid-State Circuits*, vol. 38, no. 2, pp. 263–273, Feb 2003.

[106] S. Yamauchi, T. Watanabe, and Y. Ohtsuka, "Ring oscillator and pulse phase difference encoding circuit," Patent US 5 416 444, 1995.

[107] M. Z. Straayer and M. H. Perrott, "A multi-path gated ring oscillator TDC with first-order noise shaping," *IEEE Journal of Solid-State Circuits*, vol. 44, no. 4, pp. 1089–1098, April 2009.

[108] T. Watanabe, T. Mizuno, and Y. Makino, "An all-digital analog-to-digital converter with 12-μV/LSB using moving-average filtering," *IEEE Journal of Solid-State Circuits*, vol. 38, no. 1, pp. 120–125, Jan 2003.

[109] W. C. Black and D. A. Hodges, "Time interleaved converter arrays," *IEEE Journal of Solid-State Circuits*, vol. 15, no. 6, pp. 1022–1029, Dec 1980.

[110] L. Kull, D. Luu, P. A. Francese, C. Menolfi, M. Braendli, M. Kossel, T. Morf, A. Cevrero, I. Oezkaya, H. Yueksel, and T. Toifl, "CMOS ADCs towards 100 GS/s and beyond," in *2016 IEEE Compound Semiconductor Integrated Circuit Symposium (CSICS)*, Oct 2016, pp. 1–4.

[111] J. Song, K. Ragab, X. Tang, and N. Sun, "A 10-b 800-MS/s time-interleaved SAR ADC with fast variance-based timing-skew calibration," *IEEE Journal of Solid-State Circuits*, vol. 52, no. 10, pp. 2563–2575, Oct 2017.

[112] P. Schvan, J. Bach, C. Falt, P. Flemke, R. Gibbins, Y. Greshishchev, N. Ben-Hamida, D. Pollex, J. Sitch, S. C. Wang, and J. Wolczanski,

"A 24GS/s 6b ADC in 90 nm CMOS," in *2008 IEEE International Solid-State Circuits Conference – Digest of Technical Papers*, Feb 2008, pp. 544–634.

[113] A. Petraglia and S. K. Mitra, "Analysis of mismatch effects among A/D converters in a time-interleaved waveform digitizer," *IEEE Transactions on Instrumentation and Measurement*, vol. 40, no. 5, pp. 831–835, Oct 1991.

[114] N. Kurosawa, H. Kobayashi, K. Maruyama, H. Sugawara, and K. Kobayashi, "Explicit analysis of channel mismatch effects in time-interleaved ADC systems," *IEEE Transactions on Circuits and Systems I: Fundamental Theory and Applications*, vol. 48, no. 3, pp. 261–271, Mar 2001.

[115] J. Markus, P. Deval, V. Quiquempoix, J. Silva, and G. C. Temes, "Incremental delta-sigma structures for dc measurement: an overview," in *IEEE Custom Integrated Circuits Conference 2006*, Sept 2006, pp. 41–48.

[116] J. Candy, "Decimation for sigma delta modulation," *IEEE Transactions on Communications*, vol. 34, no. 1, pp. 72–76, Jan 1986.

[117] K. Nagaraj, T. Viswanathan, K. Singhal, and J. Vlach, "Switched-capacitor circuits with reduced sensitivity to amplifier gain," *IEEE Transactions on Circuits and Systems*, vol. 34, no. 5, pp. 571–574, May 1987.

[118] B. E. Boser and B. A. Wooley, "The design of sigma-delta modulation analog-to-digital converters," *IEEE Journal of Solid-State Circuits*, vol. 23, no. 6, pp. 1298–1308, Dec 1988.

[119] T. Hayashi, Y. Inabe, K. Uchimura, and T. Kimura, "A multistage delta-sigma modulator without double integration loop," in *1986 IEEE International Solid-State Circuits Conference. Digest of Technical Papers*, vol. XXIX, Feb 1986, pp. 182–183.

[120] N. Maghari, S. Kwon, G. C. Temes, and U. Moon, "Sturdy MASHs $\Delta-\Sigma$ modulator," *Electronics Letters*, vol. 42, no. 22, pp. 1269–1270, Oct 2006.

[121] M. J. M. Pelgrom, A. C. J. Duinmaijer, and A. P. G. Welbers, "Matching properties of MOS transistors," *IEEE Journal of Solid-State Circuits*, vol. 24, no. 5, pp. 1433–1439, Oct 1989.

[122] J. Welz, I. Galton, and E. Fogleman, "Simplified logic for first-order and second-order mismatch-shaping digital-to-analog converters," *IEEE Transactions on Circuits and Systems II: Analog and Digital Signal Processing*, vol. 48, no. 11, pp. 1014–1027, Nov 2001.

[123] R. T. Baird and T. S. Fiez, "Linearity enhancement of multibit $\Delta\Sigma$ A/D and D/A converters using data weighted averaging," *IEEE Transactions on Circuits and Systems II: Analog and Digital Signal Processing*, vol. 42, no. 12, pp. 753–762, Dec 1995.

[124] H. Shibata, V. Kozlov, Z. Ji, A. Ganesan, H. Zhu, D. Paterson, J. Zhao, S. Patil, and S. Pavan, "A 9-GS/s 1.125-GHz BW oversampling continuous-time pipeline ADC achieving-164-dBFS/Hz NSD," *IEEE Journal of Solid-State Circuits*, vol. 52, no. 12, pp. 3219–3234, Dec 2017.

[125] S. Loeda, J. Harrison, F. Pourchet, and A. Adams, "A 10/20/30/40 MHz feedforward FIR DAC continuous-time $\Delta\Sigma$ ADC with robust blocker performance for radio receivers," *IEEE Journal of Solid-State Circuits*, vol. 51, no. 4, pp. 860–870, April 2016.

[126] S. Pavan, "Excess loop delay compensation in continuous-time delta-sigma modulators," *IEEE Transactions on Circuits and Systems II: Express Briefs*, vol. 55, no. 11, pp. 1119–1123, Nov 2008.

[127] E. J. van der Zwan and E. C. Dijkmans, "A 0.2-mW CMOS $\Sigma\Delta$ modulator for speech coding with 80 dB dynamic range," *IEEE Journal of Solid-State Circuits*, vol. 31, no. 12, pp. 1873–1880, Dec 1996.

[128] O. Oliaei and H. Aboushady, "Jitter effects in continuous-time $\Sigma\Delta$ modulators with delayed return-to-zero feedback," in *1998 IEEE International Conference on Electronics, Circuits and Systems. Surfing the Waves of Science and Technology (Cat. No.98EX196)*, vol. 1, 1998, pp. 351–354 vol.1.

[129] M. Ortmanns, F. Gerfers, and Y. Manoli, "Clock jitter insensitive continuous-time $\Sigma\Delta$ modulators," in *ICECS 2001. 8th IEEE International Conference on Electronics, Circuits and Systems (Cat. No.01EX483)*, vol. 2, Sept 2001, pp. 1049–1052 vol.2.

[130] O. Oliaei, "Sigma-delta modulator with spectrally shaped feedback," *IEEE Transactions on Circuits and Systems II: Analog and Digital Signal Processing*, vol. 50, no. 9, pp. 518–530, Sept 2003.

[131] S. Luschas and H.-S. Lee, "High-speed $\Sigma\Delta$ modulators with reduced timing jitter sensitivity," *IEEE Transactions on Circuits and Systems II: Analog and Digital Signal Processing*, vol. 49, no. 11, pp. 712–720, Nov 2002.

[132] M. Tanihata and T. Waho, "A feedback-signal shaping technique for multi-level continuous-time delta-sigma modulators with clock-jitter," in *36th International Symposium on Multiple-Valued Logic (ISMVL'06)*, May 2006, pp. 20–20.

[133] F. Adachi, K. Machida, and T. Waho, "A bandpass continuous-time $\Delta\Sigma$ modulator using a parallel-DAC to reduce jitter sensitivity," in *2009 IEEE International Symposium on Circuits and Systems*, May 2009, pp. 2261–2264.

[134] A. Buhmann, M. Keller, M. Ortmanns, F. Gerfers, and Y. Manoli, "Time-continuous delta-sigma A/D converters: From theory to practical implementation," in *2006 Advanced Signal Processing, Circuit and System Design Techniques for Communications*, May 2006, pp. 169–216.

[135] S. Dosho, "Digital calibration and correction methods for CMOS-ADCs," *IEICE Technical Report Vol.110 No.140*, pp. 21–30, Jul 2010.

[136] E. Hogenauer, "An economical class of digital filters for decimation and interpolation," *IEEE Transactions on Acoustics, Speech, and Signal Processing*, vol. 29, no. 2, pp. 155–162, Apr 1981.

[137] H. Aboushady, Y. Dumonteix, M. M. Louerat, and H. Mehrez, "Efficient polyphase decomposition of comb decimation filters in $\Sigma\Delta$ analog-to-digital converters," *IEEE Transactions on Circuits and Systems II: Analog and Digital Signal Processing*, vol. 48, no. 10, pp. 898–903, Oct 2001.

[138] M. Murozuka, K. Ikeura, F. Adachi, K. Machida, and T. Waho, "Time-interleaved polyphase decimation filter using signed-digit adders," in *2009 39th International Symposium on Multiple-Valued Logic*, May 2009, pp. 245–249.

[139] H. Inose, Y. Yasuda, and J. Murakami, "A telemetering system by code modulation – $\Delta\Sigma$ modulation," *IRE Transactions on Space Electronics and Telemetry*, vol. SET-8, no. 3, pp. 204–209, Sept 1962.

[140] M. Vogels and G. Gielen, "Architectural selection of A/D converters," in *Proceedings 2003. Design Automation Conference (IEEE Cat. No.03CH37451)*, June 2003, pp. 974–977.

[141] B. E. Jonsson, "A generic ADC FOM, converter passion blog," https://converterpassion.wordpress.com/a-generic-adc-fom/.

[142] ——, "Generic ADC FOM classes, converter passion blog," https://converterpassion.wordpress.com/generic-adc-fom-classes/.

[143] ——, "Using figures-of-merit to evaluate measured A/D-converter performance," in *2011 IMEKO IWADC & IEEE ADC Forum, Orvieto, Italy*, Jun 2011, pp. 248–253.

[144] R. H. Walden, "Analog-to-digital converter technology comparison," in *Proceedings of 1994 IEEE GaAs IC Symposium*, Oct 1994, pp. 217–219.

[145] ——, "Analog-to-digital converter survey and analysis," *IEEE Journal on Selected Areas in Communications*, vol. 17, no. 4, pp. 539–550, Apr 1999.

[146] G. Emmert, E. Navratil, H. Parzefall, and R. Rydval, "A versatile bipolar monolithic 6-bit A/D converter for 100 MHz sample frequency," *IEEE Journal of Solid-State Circuits*, vol. 15, no. 6, pp. 1030–1032, Dec 1980.

[147] S. Rabii and B. A. Wooley, "A 1.8-V digital-audio sigma-delta modulator in 0.8-μm CMOS," *IEEE Journal of Solid-State Circuits*, vol. 32, no. 6, pp. 783–796, Jun 1997.

[148] A. M. A. Ali, A. Morgan, C. Dillon, G. Patterson, S. Puckett, P. Bhoraskar, H. Dinc, M. Hensley, R. Stop, S. Bardsley, D. Lattimore, J. Bray, C. Speir, and R. Sneed, "A 16-bit 250-MS/s IF sampling pipelined ADC with background calibration," *IEEE Journal of Solid-State Circuits*, vol. 45, no. 12, pp. 2602–2612, Dec 2010.

[149] *A/D Converter Figures of Merit and Performance Trends*. [Online]. Available: https://www.youtube.com/watch?v=dlD0Jz3d594

[150] B. Le, T. W. Rondeau, J. H. Reed, and C. W. Bostian, "Analog-to-digital converters," *IEEE Signal Processing Magazine*, vol. 22, no. 6, pp. 69–77, Nov 2005.

[151] T. Sundstrom, B. Murmann, and C. Svensson, "Power dissipation bounds for high-speed Nyquist analog-to-digital converters," *IEEE Transactions on Circuits and Systems I: Regular Papers*, vol. 56, no. 3, pp. 509–518, March 2009.

[152] B. Murmann, "Energy limits in A/D converters," in *2013 IEEE Faible Tension Faible Consommation*, June 2013, pp. 1–4.

[153] F. Ueno, T. Inoue, K. Sugitani, M. Kinoshita, and Y. Ogata, "An oversampled sigma-delta A/D converter using time division multiplexed integrator," in *Proceedings of the 33rd Midwest Symposium on Circuits and Systems*, Aug 1990, pp. 748–751 vol.2.

[154] P. C. Yu and H.-S. Lee, "A 2.5-V, 12-b, 5-MSample/s pipelined CMOS ADC," *IEEE Journal of Solid-State Circuits*, vol. 31, no. 12, pp. 1854–1861, Dec 1996.

[155] J. K. Fiorenza, T. Sepke, P. Holloway, C. G. Sodini, and H. Lee, "Comparator-based switched-capacitor circuits for scaled CMOS technologies," *IEEE Journal of Solid-State Circuits*, vol. 41, no. 12, pp. 2658–2668, Dec 2006.

[156] F. Chen and B. Leung, "A 0.25 mW 13 b passive $\Sigma\Delta$ modulator for a 10 MHz IF input," in *1996 IEEE International Solid-State Circuits Conference. Digest of Technical Papers, ISSCC*, Feb 1996, pp. 58–59.

[157] C. Lin and T. Lee, "A 12-bit 210-MS/s 2-times interleaved pipelined-SAR ADC with a passive residue transfer technique," *IEEE Transactions on Circuits and Systems I: Regular Papers*, vol. 63, no. 7, pp. 929–938, July 2016.

[158] Zhijie Chen, Masaya Miyahara, and Akira Matsuzawa, "Fully passive noise shaping techniques in a charge-redistribution SAR ADC," *IEICE Transactions on Electronics Vol.E99-C No.6*, pp. 623–631, Jun 2016. [Online]. Available: http://i-scover.ieice.org/iscover/resource/ ARTICLE_TRAN_E99-C_6_623

[159] Y. Chae, M. Kwon, and G. Han, "A 0.8-μW switched-capacitor sigma-delta modulator using a class-C inverter," in *2004 IEEE International Symposium on Circuits and Systems (IEEE Cat. No.04CH37512)*, vol. 1, May 2004, pp. I–1152.

[160] Y. Chae and G. Han, "Low voltage, low power, inverter-based switched-capacitor delta-sigma modulator," *IEEE Journal of Solid-State Circuits*, vol. 44, no. 2, pp. 458–472, Feb 2009.

[161] H. Kotani, R. Yaguchi, and T. Waho, "Energy efficiency of multi-bit delta-sigma modulators using inverter-based integrators," in *2012 IEEE 42nd International Symposium on Multiple-Valued Logic*, May 2012, pp. 203–207.

[162] J. Park, Y. Hwang, and D. Jeong, "A 0.4-to-1 V voltage scalable $\Delta\Sigma$ ADC with two-step hybrid integrator for IoT sensor applications in 65-nm LP CMOS," *IEEE Transactions on Circuits and Systems II: Express Briefs*, vol. 64, no. 12, pp. 1417–1421, Dec 2017.

[163] Y. Lim and M. P. Flynn, "A 100MS/s 10.5b 2.46mW comparator-less pipeline ADC using self-biased ring amplifiers," in *2014 IEEE International Solid-State Circuits Conference Digest of Technical Papers (ISSCC)*, Feb 2014, pp. 202–203.

[164] ——, "A 100 MS/s, 10.5 bit, 2.46 mW comparator-less pipeline ADC using self-biased ring amplifiers," *IEEE Journal of Solid-State Circuits*, vol. 50, no. 10, pp. 2331–2341, Oct 2015.

[165] ——, "A 1 mW 71.5 dB SNDR 50 MS/s 13 bit fully differential ring amplifier based SAR-assisted pipeline ADC," *IEEE Journal of Solid-State Circuits*, vol. 50, no. 12, pp. 2901–2911, Dec 2015.

[166] Y. Chen, J. Wang, H. Hu, F. Ye, and J. Ren, "A time-interleaved SAR assisted pipeline ADC with bias-enhanced ring amplifier," *IEEE Transactions on Circuits and Systems II: Express Briefs*, vol. 65, no. 11, pp. 1584–1588, 2017.

[167] B. J. Hosticka, "Dynamic CMOS amplifiers," *IEEE Journal of Solid-State Circuits*, vol. 15, no. 5, pp. 881–886, Oct 1980.

[168] M. Steyaert, J. Crols, and S. Gogaert, "Switched-opamp, a technique for realising full CMOS switched-capacitor filters at very low voltages," in *ESSCIRC '93: Nineteenth European Solid-State Circuits Conference*, vol. 1, Sept 1993, pp. 178–181.

[169] B. Verbruggen, J. Craninckx, M. Kuijk, P. Wambacq, and G. V. der Plas1, "A 2.6 mW 6b 2.2GS/s 4-times interleaved fully dynamic pipelined ADC in 40 nm digital CMOS," in *2010 IEEE International Solid-State Circuits Conference – (ISSCC)*, Feb 2010, pp. 296–297.

[170] J. Lin, M. Miyahara, and A. Matsuzawa, "A 15.5 dB, wide signal swing, dynamic amplifier using a common-mode voltage detection technique," in *2011 IEEE International Symposium of Circuits and Systems (ISCAS)*, May 2011, pp. 21–24.

[171] J. Lin, D. Paik, S. Lee, M. Miyahara, and A. Matsuzawa, "An ultra-low-voltage 160 MS/s 7 bit interpolated pipeline ADC using dynamic amplifiers," *IEEE Journal of Solid-State Circuits*, vol. 50, no. 6, pp. 1399–1411, June 2015.

[172] M. A. Copeland and J. M. Rabaey, "Dynamic amplifier for m.o.s. technology," *Electronics Letters*, vol. 15, no. 10, pp. 301–302, May 1979.

[173] R. Matsushiba, H. Kotani, and T. Waho, "An energy-efficient $\Delta\Sigma$ modulator using dynamic-common-source integrators," *IEICE Transactions on Electronics*, vol. E97-C, no. 5, pp. 438–443, 2014.

[174] J. Hu, N. Dolev, and B. Murmann, "A 9.4-bit, 50-MS/s, 1.44-mW pipelined ADC using dynamic source follower residue amplification," *IEEE Journal of Solid-State Circuits*, vol. 44, no. 4, pp. 1057–1066, April 2009.

[175] R. Nguyen, C. Raynaud, A. Cathelin, and B. Murmann, "A 6.7-ENOB, 500-MS/s, 5.1-mW dynamic pipeline ADC in 65-nm SOI CMOS," in *2011 Proceedings of the ESSCIRC (ESSCIRC)*, Sept 2011, pp. 359–362.

[176] R. Yaguchi, F. Adachi, and T. Waho, "A dynamic source-follower integrator and its application to $\Delta\Sigma$ modulators," *IEICE Transactions on Electronics Vol. E94-C No. 5*, pp. 802–806, May 2011.

[177] B. P. Brandt and J. Lutsky, "A 75-mW, 10-b, 20-MSPS CMOS subranging ADC with 9.5 effective bits at Nyquist," *IEEE Journal of Solid-State Circuits*, vol. 34, no. 12, pp. 1788–1795, Dec 1999.

[178] X. Fang, V. Srinivasan, J. Wills, J. Granacki, J. LaCoss, and J. Choma, "CMOS 12 bits 50kS/s micropower SAR and dual-slope hybrid ADC," in *2009 52nd IEEE International Midwest Symposium on Circuits and Systems*, Aug 2009, pp. 180–183.

[179] O. Rajaee, T. Musah, S. Takeuchi, M. Aniya, K. Hamashita, P. Hanumolu, and U. Moon, "A 79dB 80 MHz 8X-OSR hybrid delta-sigma/pipeline ADC," in *2009 Symposium on VLSI Circuits*, June 2009, pp. 74–75.

[180] J. Li and F. Maloberti, "Pipeline of successive approximation converters with optimum power merit factor," in *9th International Conference on Electronics, Circuits and Systems*, vol. 1, 2002, pp. 17–20.

[181] W. I. Mok, P. I. Mak, U. Seng-Pan, and R. P. Martins, "A highly-linear successive-approximation front-end digitizer with built-in sample-and-hold function for pipeline/two-step ADC," in *2007 IEEE International Symposium on Circuits and Systems*, May 2007, pp. 1947–1950.

[182] S. M. Louwsma, A. J. M. van Tuijl, M. Vertregt, and B. Nauta, "A 1.35 GS/s, 10 b, 175 mW time-interleaved ad converter in 0.13 μm CMOS," *IEEE Journal of Solid-State Circuits*, vol. 43, no. 4, pp. 778–786, April 2008.

[183] M. Furuta, M. Nozawa, and T. Itakura, "A $0.06mm^2$ 8.9b ENOB 40MS/s pipelined SAR ADC in 65nm CMOS," in *2010 IEEE International Solid-State Circuits Conference – (ISSCC)*, Feb 2010, pp. 382–383.

[184] H. Huang, H. Xu, B. Elies, and Y. Chiu, "A non-interleaved 12-b 330-MS/s pipelined-SAR ADC with PVT-stabilized dynamic amplifier achieving sub-1-dB SNDR variation," *IEEE Journal of Solid-State Circuits*, vol. 52, no. 12, pp. 3235–3247, Dec 2017.

[185] C.-Y. Lin and T.-C. Lee, "A 12-bit 210-MS/s 5.3-mW pipelined-SAR ADC with a passive residue transfer technique," in *2014 Symposium on VLSI Circuits Digest of Technical Papers*, June 2014, pp. 1–2.

[186] A. Imani and M. S. Bakhtiar, "A two-stage pipelined passive charge-sharing SAR ADC," in *APCCAS 2008 – 2008 IEEE Asia Pacific Conference on Circuits and Systems*, Nov 2008, pp. 141–144.

[187] C. C. Lee and M. P. Flynn, "A 12b 50MS/s 3.5mW SAR assisted 2-stage pipeline ADC," in *2010 Symposium on VLSI Circuits*, June 2010, pp. 239–240.

[188] S. W. Sin, L. Ding, Y. Zhu, H. G. Wei, C. H. Chan, U. F. Chio, U. Seng-Pan, R. P. Martins, and F. Maloberti, "An 11b 60MS/s 2.1mW two-step time-interleaved SAR-ADC with reused S&H," in *2010 Proceedings of ESSCIRC*, Sept 2010, pp. 218–221.

[189] Y.-D. Jeon, Y.-K. Cho, J.-W. Nam, K.-D. Kim, W.-Y. Lee, K.-T. Hong, and J.-K. Kwon, "A 9.15 mW 0.22 mm^2 10b 204MS/s pipelined SAR ADC in 65 nm CMOS," in *IEEE Custom Integrated Circuits Conference 2010*, Sept 2010, pp. 1–4.

[190] Y. Zhu, C. H. Chan, S. W. Sin, U. Seng-Pan, R. P. Martins, and F. Maloberti, "A 50-fJ 10-b 160-MS/s pipelined-SAR ADC decoupled flip-around MDAC and self-embedded offset cancellation," *IEEE Journal of Solid-State Circuits*, vol. 47, no. 11, pp. 2614–2626, Nov 2012.

[191] M. Zhang, K. Noh, X. Fan, and E. Sanchez-Sinencio, "A 0.8–1.2 V 10–50 MS/s 13-bit subranging pipelined-SAR ADC using a temperature-insensitive time-based amplifier," *IEEE Journal of Solid-State Circuits*, vol. 52, no. 11, pp. 2991–3005, Nov 2017.

[192] J. A. Fredenburg and M. P. Flynn, "A 90-MS/s 11-MHz-bandwidth 62-dB SNDR noise-shaping SAR ADC," *IEEE Journal of Solid-State Circuits*, vol. 47, no. 12, pp. 2898–2904, Dec 2012.

[193] C. H. Chen, Y. Zhang, J. L. Ceballos, and G. C. Temes, "Noise-shaping SAR ADC using three capacitors," *Electronics Letters*, vol. 49, no. 3, pp. 182–184, Jan 2013.

[194] Z. Chen, M. Miyahara, and A. Matsuzawa, "A 9.35-ENOB, 14.8 fJ/conv-step fully-passive noise-shaping SAR ADC," in *2015 Symposium on VLSI Circuits (VLSI Circuits)*, June 2015, pp. C64–C65.

[195] M. Ranjbar, A. Mehrabi, O. Oliaei, and F. Carrez, "A 3.1 mW continuous-time $\Delta\Sigma$ modulator with 5-bit successive approximation quantizer for WCDMA," *IEEE Journal of Solid-State Circuits*, vol. 45, no. 8, pp. 1479–1491, Aug 2010.

[196] P. Holloway and M. Norton, "A high yield, second generation 10-bit monolithic DAC," in *1976 IEEE International Solid-State Circuits Conference. Digest of Technical Papers*, vol. XIX, Feb 1976, pp. 106–107.

[197] A. Iwata, Y. Murasaka, T. Maeda, and T. Ohmoto, "Background calibration techniques for low-power and high-speed data conversion," *IEICE Transactions on Electronics Vol.E94-C No.6*, pp. 923–929, Jun 2011.

[198] S. Dosho, "Digital calibration and correction methods for CMOS analog-to-digital converters," *IEICE Transactions on Electronics Vol.E95-C No.4*, pp. 421–431, Apr 2012.

[199] B. Murmann, "Digitally assisted data converter design," in *2013 Proceedings of the ESSCIRC (ESSCIRC)*, Sept 2013, pp. 24–31.

[200] A. Verma and B. Razavi, "A 10-bit 500-MS/s 55-mW CMOS ADC," *IEEE Journal of Solid-State Circuits*, vol. 44, no. 11, pp. 3039–3050, Nov 2009.

[201] S. Park, Y. Palaskas, and M. P. Flynn, "A 4-GS/s 4-bit flash ADC in 0.18-μm CMOS," *IEEE Journal of Solid-State Circuits*, vol. 42, no. 9, pp. 1865–1872, Sept 2007.

[202] T. Cataltepe, A. R. Kramer, L. E. Larson, G. C. Temes, and R. H. Walden, "Digitally corrected multi-bit $\Sigma\Delta$ data converters," in *IEEE International Symposium on Circuits and Systems,* May 1989, pp. 647–650, vol.1.

[203] C.-C. Huang and J.-T. Wu, "A background comparator calibration technique for flash analog-to-digital converters," *IEEE Transactions on Circuits and Systems I: Regular Papers*, vol. 52, no. 9, pp. 1732–1740, Sept 2005.

Index

Δ modulator 163
$\Delta\Sigma$ modulator 186
$\Sigma\Delta$ modulator 200
1.5-bits/stage 145
1/f noise 59
2-bit/cycle scheme 135

A
A/D converter (ADC) 1
acquisition time 45
algorithmic or cyclic A/D
    converter 141
alias 24
aliasing 24
analog-to-information
    conversion 30
anti-aliasing filter 25
aperture time 44
artificial neural networks 103
asynchronous 123
attenuation capacitor 96
attenuation resistor 92

B
background (BG)
    calibration 220
bandwidth 87
binary code 7
binary search algorithm 9, 120
bipolar junction transistors
    (BJTs) 51

bit resolution 3
bootstrap switch 55
bubble 109

C
calibration 100, 219
cascaded integrator-comb
    (CIC) filter 196
cascode 99
charge redistribution 123
charge injection 48, 70
charge-sharing 97 126
class-C amplifier 207
clock feedthrough 45, 77
closed-loop S/H circuit 52
coding 3
common centroid layout 96
common-mode voltage detector
    (CMD) 210
comparator 66
compensation 219
continuous-time (CT)
    $\Delta\Sigma$ modulators 187
correction 219
current-mode signal
    processing 91
current-starved VCO 154
current-steering D/A
    converter 99
cyclic thermometer 115

About the Author

Takao Waho received the B. S., M. S., and Ph. D. degrees in Physics from Waseda University, Tokyo, Japan, in 1973, 1975 and 1978. In 1975, he joined Musashino Electrical Communications Laboratories, Nippon Telegram and Telephone Public Corporation (now NTT), where he investigated III-V compound semiconductor device technology, including MBE growth, GaAs surface passivation, resonant-tunneling diodes (RTDs), and their circuit applications. He was also involved in Josephson tunnel-junction IC technology from 1978 to 1983. Since 1999, he has been a Professor of the Faculty of Science and Technology, Sophia University, Tokyo. His current research interest includes CMOS analog integrated circuits, such as analog-to-digital converters, delta-sigma modulators, and multi-valued logic circuits, as well as nanostructure semiconductor devices. He received the Distinctive Contribution Paper Award and Certificate of Appreciation from the IEEE Computer Society in 1996 and 2009, respectively.

Dr. Waho served as Editor of the IEICE Transactions on Electronics (Japanese Edition) and as Technical Committee Chairs of IEICE Electron Devices and IEEE Computer Society Multiple-Valued Logic. He also served as General Chairs of the Topical Workshop on Heterostructure Microelectronics (TWHM) held in Nagano, Japan, in August 2011 and of the IEEE International Symposium on Multiple-Valued Logic (ISMVL) held in Toyama, Japan, in May 2013.